MARRIAGE, FAMILY AND KINSHIP AMONG THE PAITE TRIBE OF MANIPUR

About the Author

Grace Don Nemching has done her post-graduation and M.Phil in Sociology from Jawaharlal Nehru University, New Delhi. She has submitted her Ph.D. thesis to the Centre for the Study of Social Sciences, Jawaharlal Nehru University.

Grace Don Nemching is currently a faculty in Sociology in Jesus and Mary College, University of Delhi. Earlier she has worked as a researcher in Indian Social Institute, New Delhi and as a guest lecturer in RAK College of Nursing, University of Delhi. Also, she is President of the Paite student body known as Siamsinpawlpi.

MARRIAGE, FAMILY AND KINSHIP AMONG THE PAITE TRIBE OF MANIPUR

Grace Don Nemching

CONCEPT PUBLISHING COMPANY, NEW DELHI-110059

Cataloging in Publication Data—DK
Courtesy: D.K. Agencies (P) Ltd. <docinfo@dkagencies.com>

Nemching, Grace Don.
Marriage, family and kinship among the Paite tribe of Manipur / Grace Don Nemching.
p. cm.
Includes bibliographical references (p.)
Includes index.
ISBN 13: 9788180695087
ISBN 10: 8180695085

1. Paite (Asian people)—India—Manipur—Social life and customs. 2. Paite (Asian people)—Kinship—India—Manipur. 3. Paite (Asian people)—India—Manipur—Religion. 4. Paite (Asian people)—India—Manipur—History. I. Title.

DDC 305.800 954 17 22

ISBN-13: 978-81-8069-508-7
ISBN-10: 81-8069-508-5

First Published 2008

Published and Printed by
Ashok Kumar Mittal
Concept Publishing Company
A/15-16, Commercial Block, Mohan Garden
New Delhi (INDIA)

Phones : 25351460, 25351794
Fax : 091-11-25357103
Email : publishing@conceptpub.com

Preface

This is a descriptive study of the Paite tribe based on the materials taken from secondary sources like published or unpublished, books various journals and published articles.

The focus is on Marriage, family and kinship which is crucial to understand change and continuity in the life of the Paite tribe.

This book focuses mainly on the *micro-structures*, that is, structures that have a narrow extension of relational linkage; these relations may be confined to a linguistic region or even a far narrower territorial limit based on kinship ties. Its examples are: the caste, family and village communities.*

This book is divided into six chapters as follows: 1st Chapter is the *Introduction*. Here I discuss about the geographical features of *Manipur,* how did the term *'Paite'* came about. The origin and historical account of the *Paite* and also the *Focus* and *Objectives* of my study. 2nd Chapter is devoted to theoretical orientations, what are the various theories regarding *Family, Marriage* and *Kinship.* Also, there are definitions and explanation of the term *'Tribe'* and its features. 3rd Chapter is the central theme of this book which discusses the system of *Marriage, Family* and *Kinship* among the Paite. It contains various existing customs and traditions surrounding these three central themes. 4th Chapter is regarding political organization and social control, wherein I dealt about the *traditional-political structures* and *way of social control,* to contextualise the concept of marriage, family and kinship. 5th Chapter is about *Religion,* which will mainly focus on *Christianity,* as majority of the *Paite* are Christians, the account of how Christianity came to the *Paite,* what changes it brought about, the work done by the *Christian Missionaries* and also focus on changes in family, marriage and kinship in recent times. 6th Chapter is the *Conclusion.*

---

* Yogendra Singh, *Modernization of Indian Tradition* (1972).

Contents

1

Introduction

Manipur literally means the "Land of Gems" and has been referred to by various connotations like "Jewel of India", "Kashmir of Eastern India", as the "Switzerland of India" as depicted by Lord Irwin.

Manipur is a small hilly state in North-Eastern region. Its total area is 22,356 sq.km. Manipur is bounded by Nagaland in the north, Mizoram in the south, Myanmar in the east and Cachar district of Assam in the west (Fig. 1.1).

Manipur came under the British Rule in 1891. It was merged with the Indian Union in October 1949. Manipur became a full-fledged 'State of India' on the 21st January, 1972 as the 20th state of the Union.[1]

Manipur is sharply divided into two cultural zones, the valley and the hills. The climate of Manipur is pleasant and healthy. The main crops in the state are paddy, maize, yam, millets. The hills which cover nine-tenths of the total area are inhabited by 33 tribes and the valley by the Meiteis, who are in majority.[2]

All the tribes of Manipur can be roughly placed under two sections—the Kuki-Chin-Mizo and the Nagas. The term Kuki-Chin-Mizo covers a large number of tribes and sub-tribes other than the Nagas.

Manipur, is essentially[3] an agrarian state and the population is mostly rural in character. Cultivation of crops is the dominant economic activity in Manipur. The main method used is *Jhum cultivation or slash/burn cultivation.*

Manipur is essentially a mountainous state. Physiographically, it can be divided into three well defined regions:

1. The Manipur Valley.

2. The Manipur Hills comprising about 91 per cent of the state area.
3. The Barak plains, on the western border of the state is the smallest physiographic region of Manipur state, it comprises only about 1 per cent of the state area.

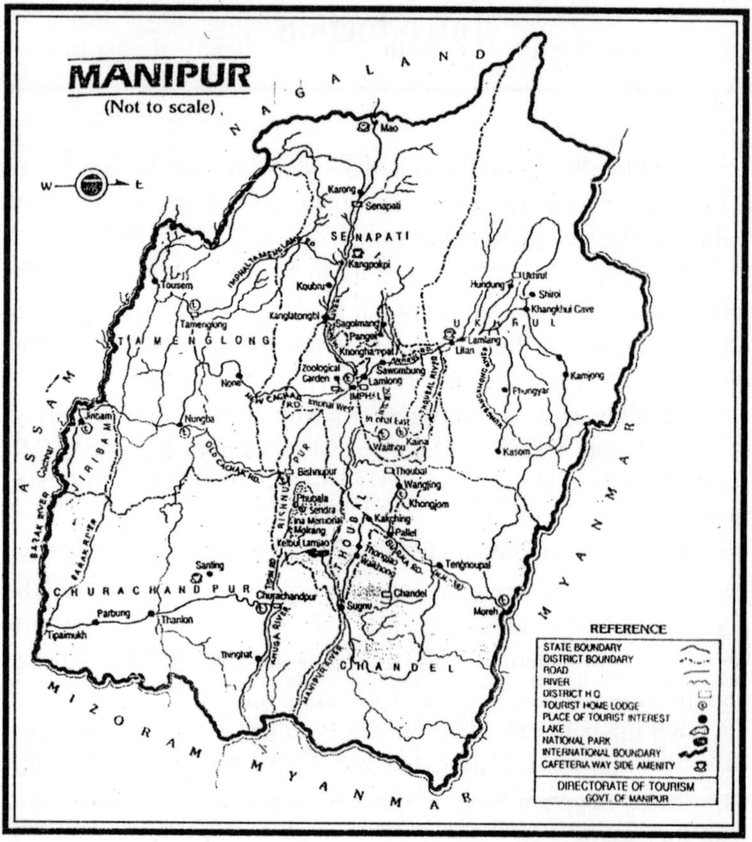

Fig. 1.1

At present the total population of Manipur is about 18,37,149 persons. Out of which the Tribal community constitute nearly 34.4 per cent, *i.e.* 6,32,173 persons out of which the *Paite tribe* constituted 2.22 per cent which is around 40,792 persons out of the total population.[4]

The *Paite Tribe* constitutes the fifth largest tribe in Manipur, out of the 33 tribes recognised as Scheduled Tribes by the Government of India.

The origin of the word 'Kuki' is not known,[5] Horato Bicker Staffe Rowney writing on the wild tribes of India with a few paragraphs on 'Kookies' records that the application 'Kookie' is equally unknown and given to them by inhabitants of East Bengal (now Bangladesh), they are known by different names at different places. In Burma (now Myanmar) they are known as 'Chin'. In India and Chittagong (in Bangladesh) they are known as 'Kukis'.

According to Sir Major General James Johnstone, he talks of the Kukis as a wandering race consisting of several tribes. They were first heard of as Kukis in Manipur between 1830 and 1840.[6]

According to popular mythology[7], the Kukis believed that they have emerged from 'Khul' the sub-terranian of the Earth. S. Bartaki has noted that the Kukis originally belonged to a place called *Sinlung* in South-West China.

Their Chinese origin may also be supported by their musical instruments common with the Chinese. A great famine (according to some, a great flood) forced them to leave China and to search for a new land down the Chindwin valley to the Chin hills of Burma (now Myanmar).[8]

Their salient characteristics were accorded by Bertram S. Carey and H.N. Tuck as the slow speech, the serious manner, the respect for birth and the knowledge of pedigrees, the duty of revenge, the taste for and the trecherous method of warfare, the curse of drink, the virtue of hospitality, the clannish feeling, the vice of avarice, the filthy state of body, mutual distrust, impatience under control, arrogance in victory, speedy discouragement and panic in defeat.[9]

Sir James Johnstone in his observation, recollected that the Kukis are strictly monarchical, and their chiefs are absolutely despotic and may murder or sell their subjects into slavery without a murmur of dissent.[10]

Some writers were aware of the fact that the people whom they called by different names such as Kuki, Chin etc. were but the same. Carey and Tuck mentioned that the Kuki of Manipur, the Lushai Hills of Bengal and Assam and the Chin originally lived in what we know as Thibet (Tibet) and are one and the same stock; the form of government, method of cultivation, manners and customs, beliefs and tradition all point to one origin.[11]

The Kukis belong to the "Burmese-Kuki-Chin-Kachin-Lolo" group of Tibeto-Burman Mongoloids. The tribes belonging to the Kuki group are believed to be descendants of one common ancestor. The belief that the common ancestor they share is called 'Man Ma Shi Nao' which means mortal. All the Kuki-Chin tribes lived in a nomadic state for some centuries and gradually they migrated to places like Manipur, Burma (now Myanmar), Mizoram etc.

In 1956, under the list of tribes recognised and scheduled by the Government of India in Manipur, the tribes who were known as 'Kukis' petitioned to be known according to their own distinct tribe name, which was granted to them, so in all 33 tribes, with their distinct tribe names came into existence, the Paite tribe was one of them. The Paites are the fifth largest tribe in Manipur (Fig. 1.2).

The name Paite, according to M.C. Goswami and H. Kamkhenthang (1972 : 21), comes from two words Pai and Te which refers to 'people moving towards different places'.

The Paites inhabit mainly the Churachandpur District of Manipur. They speak Paite dialect which belongs to the Tibeto-Burman family of Kuki-Chin group.[12]

The Paites are a distinct tribe having their own customs and traditions and rules regarding succession, marriage, discent etc.

Prof. J. Roy[13] wrote about the Paite, observing it is interesting to note that, the meaning of Paite, one of the Kuki-Chin tribes is a people in the process of going or immigration. They are Tibeto-Burmese in speech and Mongoloid in physical characteristics.

A one time noted diplomat of Burma, U Vum Ko Hau identified the Kuki-Chin people as follows:

The Chin Hills Gazetteer recorded the fact that Zo (Chins) and the so called Kukis (among them the Paite tribe) were one and the same race.

The Paite according to their traditional story originated either in Tibet or in China from where they entered India via Burma (Fig. 1.3).

Hill Tribes of Manipur

Naga Tribe

Tribe	Distribution
(A) Tangkhuls (Ukhrul Sub-division adjacent to Burma)	
(B) Zemis allied (Temenglong sub-division in the Cachar Naga and Western Hills or Manipur)	
(C) Rongmeis (Kabuis) Tribe	
(D) Mao	Live in the northern sub-division of Mao and Sadar Hills bordering Nagaland
(E) Maram	
(F) Thangals	
(G) Maring	
(H) Moyon	Occupy the Tengnoupal Sub-division
(I) Monsang	On the south-west bordering Burma (formerly belong to old Kuki clan) but have changed their affiliation and now includes in the Naga-subnation
(J) Lamgang	
(K) Anal	
(L) Maram	

The Non-Naga (Kuki Chin) Family

Tribe	Distribution
(A) Aimei	
(B) Chiru	
(C) Purum	Ethnologically they belong to the Kuki-Chin family All these tribes living in the Sadar hills. Churachandpur. Jiribam and Tengnoupal sub-division of Manipur
(D) Paite	
(E) Thadou	
(F) Chothe	
(G) Kom	
(H) Gangte	
(I) Simte	
(J) Vaiphei	
(K) Hmar	
(L) Zou	
(M) Ralti	
(N) Any Mizo	

Source: *History of Manipur,* Roy (1973).

Fig. 1.2

Moreover, Grierson stated that the Zahao or Yahow and Wheno, the two most powerful communities of Tashons in the Chin Hills were formally known as Pois, Poi-Te and Paite. Another authority on tribes, Fuchs supported Grierson when he said from A.D. 1406-1700 the Paite inhabited the Chin Hills.[14]

Map of Chin State of Burma

Fig. 1.3

In his article "Chin" in *Encyclopaedia Britannica,* Christopher Von Haimendorf recounts the stories that "white Chins came from a man and woman that fell from the clouds." How and when did the *Paite* come to Manipur is still obscure. It seems there were times when Manipur experienced several foreign invasions.

Prof. J. Roy writes: "the fertile valley of Manipur witnessed the invasion of different tribes from time immemorial. At different periods, the Nagas, the Kukis, the Shans, the Chinese came and settled in the land and merged themselves into the Manipur community."[15]

The name and the tribe Paite was not officially known in the days the British were in control of Manipur, as the Paite tribe where clubbed together with other tribes under the term "Kukis". Hence, the Paite people became restless, since, the attainment of Independence by India on 15th August, 1947, for the recognition of their tribe as a distinct one, which finally they succeeded in attaining in 1957, when the term "Paite" was included in the Scheduled Tribes list of Manipur released by the Government of India.

The *Chin*[16] have much in common with the *Kuki, Lushai,* and *Lakher* people and speak *Tibeto-Burmese* languages. At some places, the *'Chin'* is used to refer to the *Kuki tribes. Chin,* a group of Tribes of Mongol origin, occupying the southern most part of the mountain ranges separating Burma from India. Their History from the 17th to the late 19th century was a long sequence of tribal wars and feuds.

The first British expedition into the Chin Hills in 1889 was soon followed by annexation and British administration ended raids by the Chin on the Burmese plains. Chin villages, often of several hundred houses, were traditionally self-contained units, some ruled by Council of Elders, others by Headmen. There were also hereditary Chiefs who exercised political control over large areas and received tribute from cultivators of the soil.

Agriculture is the basis of Chin Economy, land is cultivated in rotation, consecutive for several years being followed by reversion to forest. Rice, Millet and Corn (maize) are the main crops. Domestic animals are kept for meat not for milk nor used for traction. Chief among them is the Mithun, a domesticated breed of Indian wild ox.

Prowess in Hunting has religious significance of the Chin, the slayer of much game is believed to enjoy high rank in after life. Status in life and presumably in, after life, is achieved by providing feasts. They practice *Polygyny* and trace their descent through the paternal line. Young people are expected to marry outside the Paternal clan. Traditional religion comprises a belief in numerous dieties and spirits, which may be propitiated by offerings and sacrifices. Christian missions have made many converts. The tribes have retained their Identity, however and Burmese influence has remained limited.[17]

Objective of Study

My objectives to study the Paite tribe are as follows:

First, I belong to the Paite tribe, no books have been published or study conducted by a female sociologist about this tribe.

Second, as I belong to this tribe it will give me the advantage of collecting information and conducting interviews, if necessary in the Paite dialect.

Third, having studied in Delhi for more than 20 years, it will give me an opportunity to go back to my roots and know more about my tribe.

Focus of the Study

The main focus will be on the study of Family, Marriage and Kinship and also Religion which has played a very significant role among the Paite tribe. There is a description of the Paite's place of origin, from where they have come to settle down, and how they got their name. The Paite's Kinship relations, rules of descent, marriage, familial relations, Kinship terminology and also political organisation and social control are discussed here.

Methodology

Personal documents would be used, these documents would be in the form of solicited personal documents, *i.e.* documents that are produced at the request of the researcher and unsolicited personal documents that are created by the subject either for his or her use *e.g.* a personal diary etc.

Published and unpublished written materials would be used.

If necessary, Unstructured Open Ended Interview of the elder members of the Paite tribe would be taken up, this type of interview would be useful in getting an indepth insight to a person's experiences, lifestyles, the various stages of his life etc.

NOTES

1. B.K. Ahluwalia, *'Social Change in Manipur'*, (1984).
2. T. Liankhohau, *'Paite Tribe—A Transformed Community'*, (1994).
3. *Encyclopaedia of India,* Vol. XXI, pp. 1-3, (Manipur).
4. *Economic and Political Weekly,* Sept. 29, 2001. Vol. XXXVI pp. 3674-3677.
5. James Johnstone, *Manipur and the Naga Hills,* (1876).
6. *Ibid.*
7. Ashok Kumar Ray, *Study of Thadou-Kukis in Manipur,* (1990).
8. *Ibid.*
9. Charey, S. Bertham and Tuck, H.N. *The Chin Hills,* Vol. 1, (1896).
10. Sir James Johnstone, *op. cit.,* no. 3.
11. *Ibid.*
12. Sipra Sen, *Tribes and Castes of Manipur,* (1992).
13. T. Liankhohau, *op. cit.,* no. 2.
14. *Ibid.*
15. In T. Liankhohau, *op. cit.,* no. 2. p. 10.
16. Encyclopaedia Britannica, Vol. V, p. 555.
17. *Ibid.*

2

Theoretical Orientations

The Paite tribe is a tribe belonging to the Kuki-Chin-Mizo group of the Tibeto-Burman family.[1] The Paite tribe mainly inhabit the hilly parts of Manipur and they are the 5th largest tribe in Manipur, out of the 33 tribes recognised by the Government of India as Scheduled Tribes in Manipur.

The word *'Pai'* means to go and *'Te'* is a suffix signifying the plural number of a third person. The Paites were known by the term 'Kuki' which was a nomenclature for the other 33 tribes as well in Manipur, but the separate designation 'Paite' came into existence in 1956-57 when the Government of India came out with a list of Scheduled Tribes, which also accorded separate terms to the other 33 tribes as well. The word Tribe is derived from the Greek word 'Tribuz'.

Meaning

The modern use of the word 'tribe' started by the early explorers who found backward human communities living in uncivilised manner.

The English word 'Tribe'[2] according to Reader's Digest Encyclopaedia Dictionary is used in 5 different senses, these are:

> *First, Group of primitive* or *barbarous clan* under recognised chiefs.
>
> *Second,* meaning is related to Roman history (clans of Romans were termed as tribes).
>
> *Third,* the word also refers to similar division whether of natural or political.

Fourth, meaning relates to unit of taxonomy.
The *Fifth* meaning is used for 'large numbers'.[3]

The Dictionary of Anthropology (1956) describes tribe: A social group, usually with a definite area, dialect, cultural, homogeneity and unifying social organization may include several sub-groups, such as sibs or villages.*

Rivers (1932) indicate the term tribe refers to "a social group of simple kind, the members of which speak a common dialect, have a single government and act together for some common purpose like warfare".*

Das (1953) defines the term tribe as: A tribe generally has a common name, a common habitat, a common language, a common culture and a feeling of unity among its members as against members of other tribes. These common features give the tribe its individuality and thereby keep it distinct and different from the other tribes as advanced groups.*

Majumdar (1967) defined 'tribe' as: collection of families or group of families which occupy the same territory, speak the same language and observe certain taboos regarding marriage, profession, or occupation and have developed well assessed reciprocity and mutuality and obligations.*

The term tribe commonly signifies a group of people speaking a common language, observing uniform rules of social organisation and working together for common purposes.[4]

Elwin (Verrier) had distinguished three sections of tribes with respect to their degree of Hinduization:

> *First,* such sections as the Rajgonds and others who have successfully fought the battle and are recognised as members of a family of high status within Hindu society.
> *Secondly,* the large mass that has particularly Hinduized and has come into close contact with Hindus, and
> *Thirdly,* the hill sections, which have exhibited the greatest power of resistance to the alien cultures that have pressed their border.[5]

* In Man in India, *March* (1991) 71(1), pp. 151 and 152.

The Paite Tribe would belong to the third category, because of their almost isolated and autonomous existence having their own customary laws, rites, rituals, customs still in practice and their geographical isolation is because of the hilly terrain which makes it difficult to penetrate.

Usage

The term 'tribe' is not precisely defined in the Constitution of India, which however refers to such communities as Scheduled Tribes.

Professor D.N. Majumdar has ably said, 'all tribes in India are in transition. Just as in case of language, where no sharp geographical line can be drawn, but where the transition from one dialect to another can best be represented, not by a line, but by colours which fade into one another, so also in the case of tribal cultures, it is hard to say when a culture ceases to be 'Tribal'.[6]

An important issue[7] in the study of tribes today is how we understand the relationships between tribe and civilization. It is here that, the author makes a distinction between two approaches: "Evolutionary and Historical".

Evolutionary Approach

The evolutionary approach takes a long range view of the passage of time and stresses the succession of social formations.

Historical Approach

The historical approach limits itself to a particular framework of space and time and stresses the co-existence of different social formation within that framework.

.The present list shows more than *400 tribes* with an aggregate population of over *fifty million persons,* accounting for 7.76 per cent of the total population. Even a cursory look will show that it includes the widest variety of social formations from small food gathering bands to vast populations of settled agriculturists comprising three million persons and more.

Caste and Tribe emphasize and perpetuate collective identities in strinkingly similar ways; a caste or tribe may change its name and within limits also its mode of livelihood and yet retain its collective identity.[8]

Defining *'tribe'*[9] has conceptual as well as empirical problems for the academician. This term of administrative convenience has now been adopted by the tribals themselves to mean, the dispossessed, deprived people of a region. There is no claim to being the original inhabitants of that region, but a prior claim to the natural resources is asserted *vis-à-vis* the outsiders and the dominant caste. The tribal identity now gives the marginalised people self-esteem and pride.

The idea of 'indigenous people' is an issue of considerable contention in India. This was hardly so till a few years ago. In fact, social workers, administrators, politicians and even scholars widely used the term to refer to a certain category of people. They hardly felt any unease in the use of native equivalent of this term, *viz., adivasi.* Ghurye had, of course, some reservation to the use of such terms; the expression he used was *'so-called aborigines'*.

Again, it is they who took the term along with all the prejudices and conjectures to the masses. That is how the identity of *adivasis* has entered into the consciousness of the tribal people. The identity that was forced upon them from outside precisely to mark out differences from the dominant community has now been internalised by the people themselves. Not only has it become an important mark of social differentiation and identity assertion but also an important tool of articulation for empowerment.[10]

In regard to the study of a group the following characteristics should be noted:

Local or *territorial limitations*[11], if any composition [Demography].

Mode of acquiring and losing membership, e.g. by birth, adoption and marriage etc.

Form, i.e. the constitution and forms of behaviour distinguishing it from other groups.

Function: the relation of its activities to those of other groups and of the total community.[12]

In all societies[13] people are bound together in groups by various kinds of bonds. The most universal and most basic of these bonds is that which is based on reproduction, an inherent human drive and is called *Kinship.* The desire for the production give rise to two kinds of bonds:

Firstly, there is the bond between spouses and their relatives, this kind of bond arises out of a socially or legally defined marital relationship which is called affinal kinship, and the relatives so related are called affinal kin.

Secondly, there is the bond between parents and children and between siblings, *i.e.* children of the same parents. This relationship based on blood ties is called consanguineous kinship.[14]

The study of *Kinship*[15] as an aspect of social structure is replete with legal terminology and concepts: rights, claims, obligations, contract, agnation, corporate etc. The reasons for this is quite simple—inheritance, succession and marriage, every society makes some provision for the transfer of property and social position on death and transfer is usually to a kinsman. A good deal of any legal system, including our own is concerned with laying down rules about who may succeed whom, about obligations and duties. We can always look at what anthropologists have called *Kinship* relations is two ways:

On the one hand, we can look at the total society and ask how it forms its kinship groups and how they function and *on the other hand,* we can look at the network of relationships that bind individuals to each other.[16]

There are groups referred to as *Corporate* and *Kindred.*

Corporate

A group may be spoken of a 'corporate' when it possesses anyone of a certain number of characters. If its members, or its adult male members, or a considered proportion of them come together occasionally to carry out some collective action, *e.g.* the performance of rites; if it has a chief or council who are regarded as acting as a whole; if it possesses or controls property which is collective, as when a clan or lineage is a land-owning group. In parts of Africa, it is very common to find that land is held or owned by lineage groups, which are thus corporate groups.

Kindred

The totality of matrilateral and patrilateral kin recognized by a person within a certain degree is called kindred. In this an individual can trace a geneological tie through both of his parents in a system of reciprocal claims and obligations. Kindred comprises all stocks of which an individual is a member to a certain degree.

Marriage

Marriage had been discussed largely in the context of recruitment to kinship groups; legitimate marriage was necessary to provide for legitimate offspring to replenish the group.

Levi-Sirauss, argued, that kinship groups were simply units in a system of 'alliances' made or 'expressed' by marriage. The real differences between kinship system, then lay in the different way in which they moved around the system in marriage.[17]

In ordinary English usage, the word 'marriage' is used in at least four distinguishable but overlapping senses to refer to: The legal, rights and duties *vis-à-vis* each other of 'husband' and 'wife' on the one hand and of 'wife's husband' and 'wife's child' on the other. Hence, marriage provides the children of a woman with a legitimate status in society. The practical arrangements by which husband, wife and children combine to form a household. The ceremonial [wedding] through which the husband and the wife are put into legally enforceable contractual relations with one another in the first place. The relationship of alliance which links the two affinally associated 'families' which are represented in the persons of the husband and wife.[18]

Marriage, ensures a biological satisfaction (that of sex) and a psychological satisfaction (that of having children on the individual plane). On the wider collective plane, it ensures a two-fold survival, *viz.* that of the group and its culture.[19]

G.P. Murdock,[20] emphasises *marriage* as living together with regular sexual relationship and economic co-operation.

Harry M. Johnshon, believes that *marriage* is a stable relationship which society allows to man and woman in the community without losing its existence.

The study of Kinship and Marriage can be looked at as an arrangement which enables person to live together and co-operate with one another in an orderly social life. The term *Consanguinity* is sometimes used as an equivalent of *Kinship* but the word has implications which must be avoided.

Consanguinity refers properly to a physical relationship, but in kinship we have to deal with a specifically social relationship.

Kinship

Kinship, therefore, results from the recognition of a social relationship between parents and child which is not the same thing as the physical relation and may or may not coincide with it. The reality of a kinship system as a part of a social structure consists of the actual social relations of person to person as exhibited in their interactions and their behaviour in respect of one another. A kinship system thus presents to us a complex set of norms, of usages, of patterns of behaviour between kindred.

Two persons who are kin are related in either of the two ways: either one is descended from the other, or they are both descended from a common ancestor.

One principle that may be adopted is the *simple cognatic kinship:* to define the kin of a given person, his descent is traced back a certain number of generations, to his four grandparents, his eight great-grandparents, or still further, and all descendants of his recognized ancestors through both females and males are his cognates.

Another way of ordering the kindred may be illustrated by the system of ancient Rome. Within the body of a person's recognized cognates certain are distinguished agnates. Cognates are agnates if they are descendants by male links from the same male ancestor.

In some other societies, there is similar emphasis on unilineal descent through females. With such a system a person distinguishes from the rest of his cognates those persons who are descended by female links only from the same female ancestress as himself. We can speak of these as his matrilineal kin.

For a *Kinship system,* to exist or to continue in existence it must 'work' with at least some measure of effectiveness. It must provide an integration of persons in a set of relationships within which they can interact and co-operate without too many serious conflicts.[21]

According to ***Harry M. Johnson, Kinship,***[22] has six important bases:

- **(i)** ***Sex*****:** The terms 'brother' and 'sister' also indicate the sex of blood relations.
- **(ii)** ***Generation*****:** The terms 'father' and 'son' indicate two generations on one hand and on the other blood relation also.
- **(iii)** ***Closeness*****:** The relationship with son-in-law and father's sister's husband is based on closeness but these relations are not based on blood relations.
- **(iv)** ***Blood Relation*****:** The kinship based on blood relation is divided into lineage such as grandfather, father, son, grandson, etc. and also such as uncles and their children.
- **(v)** ***Division*****:** All kinship relations are generally divided into two branches, *e.g.* father's father (paternal grandfather) and mother's father (maternal grandfather) brother's daughter and sister's daughter, son's son and daughter's son.
- **(vi)** ***Binding Thread*****:** The binding thread of these relations is close on blood relationships.

Marriages

Marriages, births, deaths or initiation at puberty are re-arrangements of structure that are constantly recurring in any society; they are moments of the continuing social process regulated by customs; they are institutionalized ways of dealing with such events.

The rules relating to marriages between related persons in any society are an intrinsic part of the complex of institutions that make up a kinship system. In each instance the rules can only be understood by reference to the systems to which they belong and the social structures that constitute the basis of the system.

By a *Marriage,* certain existing relationships particularly, in most societies, those of the bride and her family is changed. New social relations are created, not only between the husband and wife and between the husband and the wife's relatives on the one side and between the wife and husband's relatives on the other but also, in a great many societies; between the relatives of the husband and those of the wife, who, on the two sides are interested in the general law which each instance exemplifies is that the rules have for their function to maintain the continuity of the general system of institutional marriages, either by preventing marriages which would be disruptive or by encouraging marriages which reinforces the existing re-arrangement of persons.[23]

Marriage,[24] is a socially recognised union between persons of opposite sex. It derives its importance as a social institution from the fact that it provides a stable foundation for the creation and organization of a conjugal group. Particularly all societies consider married life the most normal and desirable type of existence of adults.

The spouses are expected to find in such relationships not merely regular satisfaction of sexual needs and co-operation in economic matters but emotional response as well. Most societies look upon marriage as a legal contract either between individuals or between their respective families.[25]

Kinship and Marriage

Kinship and Marriage are the basic facts of life. They are about birth and copulation and death. The study of kinship and marriage can perhaps be reduced to four basic 'Principles':

(i) The women bore the children.
(ii) The men impregnate the women.
(iii) The men usually exercise control.
(iv) Primary kin do not mate with each other.[26]

Many sociologists have regarded the family as the cornerstone of society. It forms the basic unit of social organization and it is difficult to imagine how human society could function without it.[27]

Talcott Parsons[28] concentrates his analysis on the family in modern American society. However, his ideas have a more general application since he argues that the American family retains two 'basic and irreducible functions' which are common to the family in all societies. These are the *Primary Socialization of Children and Stabilization of the Adult Personalities* of the population of the society.

All societies recognise the existence of certain close knit internally organized co-operative units intermediate between the individual and the total society of which he is a part.

Theoretically, every person is assigned to one or the other of these units on the basis of biological relationships established through mating or common ancestry. Such units always have specific functions with relation both to their members and to the total society. Membership in the unit entails upon the individual specific rights and duties with regard to other members and also a series of rather clearly defined attitudes.

The unit is expected to be the primary focus of the member's interests and loyalties. Those who belong to it are duty bound to co-operate with and assist each other and to place each other's interests above the personalities within the unit is close and continuous.

Some socially ascribed functions of the family unit are:

(a) Economic production (specialised roles for males and females).
(b) The care of the aged.
(c) Protecting its member's interests against outsiders.[29]

The family is a social group characterised by a common residence, economic, co-operation and reproduction.

Three distinct types of family are mentioned by ***Murdock:***

> *First* and the most logic is called the *Nuclear Family* which consists of a married man and woman with their offspring.
> *Second,* Polygamous Family which consists of two or more nuclear families affiliated by plural marriages, *i.e.* having one married parent in common.
> *Third,* Extended Family which consists of two or more nuclear families affiliated through an extension of the parent-child relationship rather than of the husband-wife relationship, *i.e.* by joining the nuclear family or a married adult to that of his parents.

Whatever larger familial forms may exist, and to whatever extent the greater unit may assume, family has its vital functions, sexual, economic, reproductive and educational.[30]

In the early 1950s[31] there was an academic debate which took place to determine whether the term 'family' and 'marriage' represented 'universal' human institutions.

Those who gave positive answer arrived at the following propositions: Human children, if they are to survive and become adults must be cared for by adults at least until adolescence. Because of this biological necessity all human societies must have rules which establish the legitimacy of young children as members of a caring domestic unit. The status of legitimacy puts obligations on the members of the domestic unit to provide care for the child. Marriage is the institution which establishes the status of legitimacy.

In ordinary circumstances, in the principal English speaking societies, this caring environment is provided by a domestic household, the core membership of which is a single nuclear family of married parents and children.

Prior to the 19th century, the term *'family'* was used in just this way to denote the member of a household, especially the servants, who is the more affluent social classes, always greatly outnumbered the inner kin-group of parents and children.

An alternative usage, which goes back at least to the *17th century,* made *'family'* a widely dispersed group of relatives, loosely linked by ties of *'blood'* and '*affinity*' but not necessarily associated with anyone household. A family then consists of those descended or claiming descent from a common ancestor.[32]

One of the few widely accepted generalizations of social science asserts that the *family* is an institution found in all societies. Because they are a part of all human experience and such a powerful enency in the formation of adult attitudes, familial relations have always been a subject of interest and discussions.[33]

It is to be remembered[34] that *descent* here refers to the social relationship of parents and children, not to the physical relation. Kinship is thus based on descent, and what just determines the character of a kinship system is the way in which descent is recognized and reckoned.

One important way in which the unilincal descent principle may be used is in the formation of recognized lineage groups as part of the social structure, an agnatic lineage consists of an original male ancestor and all his descendants through males of three, four, five or a generation.

A women belongs to the lineage of her father but her children do not. Lineage groups may be agnatic (male ancestor) or matrilineal (female ancestor). In matrilineal system, a man belongs to his mother's lineage but his children do not.

The *Paites* trace their descent through Agnatic descendants *i.e.* from a male ancestor.

The contributions of *Levi-Strauss* to the study of kinship systems are of a different character, the principle of reciprocity, as manifested in various froms of exchange in social life, is central to *Levi-Strauss* view of social institutions.

Kinship in human society is established and perpetuated through specific forms of marriage and marriage as a form of exchange involves the circulation of women. He is, therefore, particularly concerned with what he calls "elementary structures" or those characterized by preferential marriage with a particular category of kin, usually a *Cross-Cousin.*[35]

NOTES

1. G.A. Grieson, *Languistic Survey of India,* Vol. III, Part 3, (1903).
2. *Ibid.*
3. S.K. Tiwari, *Antiquity of Indian Tribes,* (1998).
4. *The New Encyclopaedia Britanica,* Vol. X, p. 115.
5. *Ibid.*
6. N.K. Bose, *Culture and Society in India,* 1977, p. 191.
7. Beteille, Andre, *Society and Politics in India,* 1991.
8. *Ibid.*
9. *Ibid.*
10. Virginius XAXA In *EPW* December 18, 1999, pp. 3589-3595.
11. A.L. Epstein, *The Craft of Social Anthropology,* (1967).
12. *Ibid.*
13. D.N. Majumdar and T.N. Madan, *An Introduction to Social Anthropology,* (1967).
14. *Ibid.*
15. Robin Fox, *Kinship and Marriage—An Anthropologist Perspective,* 1987.
16. *Ibid.*
17. Levi-Strauss, *Elementary Structure of Kinship,* 1949.
18. Edmund Leach, *Social Anthropology,* (1982).
19. D.N. Majumdar and T.N. Madan, *op. cit.* no. 13.
20. A.R. Radcliffe Brown and Daryli Fordes, *African Systems of Kinship and Marriage,* (1967), (Edited).
21. *Ibid.*
22. A.R. Radcliffe Brown and Daryll Fordes, *op. cit.* no. 11.
23. *Ibid.*
24. *Ibid.*
25. Linton, R. *The Study of Man (An Introduction)* (1936).
26. Robin Fox, *op. cit.* no. 2.
27. M. Hara Lambos with H.M. Head, *Sociology: Themes and Perspectives,* (1980).
28. Linton, R., *op. cit.* no. 13.
29. *Ibid.*
30. Murdock, *Social Structure,* (1965).
31. Edmund Leach, *op. cit.* no. 9.
32. *Ibid.*
33. *International Encyclopaedia of Social Sciences,* pp. 301-322, Vols. V and VI.
34. A.R. Radcliffe Brown and Daryll Fordes, *op. cit.* no. 11.
35. *Ibid.*

3

Marriage, Family and Kinship

This chapter deals with the central theme of this book which is family, marriage and kinship among the Paite tribe. It will be a descriptive analysis and there will be Paite terms included alongside to describe the various events taking place.

Marriage

Procedure for Engagement (Mou Ki-Hel Dan)

When a boy[1] is intending to get married, the boy's family will send a messenger to the girl's house if they can come and offer tea, when it is accepted, the parents of the boy seeking a bride brings a pot of *'Zu'* (before rice beer but now tea) to the parents of the girl and start negotiation; If agreeable to both the parties, a date has to be fixed for marriage; if not agreeable then the parents of the girl have to return *Zu/tea* by offering their own *Zu/tea.*

After making an agreement between a boy and a girl for marriage, if any boy or girl other than the concern parties spoils the engagement, it is said to be *Zawl Thu Kal A Sial Thawl* which means to cut off the rope which ties *Mithun* the offender is liable to pay a fine of one *Mithun* and *Zubel**/*Tea.*[2]

Mithun is a cattle, it plays an important role in the life of the *Paite* tribe, it is the basis for family, marriage and kinship ties. This reminds us of the Nuer tribe of Sudan, written about by Evans-Pritchard, their whole life revolve around the care of cattle and fights would take place regarding their cattles and disputes and

* Zubel means Rice Beer.

other arrangements would be settled by exchanging cattles. The cattle breeding was one of the most important economic activities, a wife would be bought by giving the girl's family cattles as demanded.

In many African 'Marriage' institutions in their traditional form, the husband and his kin made substantial payments in cattle to the male kin of the bride. The Payment was a symbolic expression of the legitimacy of the marriage and of the value that was attached to it. The higher the payment, the greater is the honour to the Bride.

Now let us go back to the Paite.

Marriage (Kitenna)

If there is[2] no breach of engagement between a boy or a girl till the date fixed for marriage, a marriage ceremony is performed with the knowledge of the *Household Council (Indongta)* (details on page 37) which consists of family members, relatives and close friends and also with the knowledge of the public in the church, and then after, the boy and girl becomes husband and wife.

At the times of marriage the bride and the bridegroom respectively have a brides maid and best man standing alongside.

Brides Maid is called *Mou Liah/Moulawm.*

Best Man is called *Moupa Liah/Moupa Lawn.*

Lawichal

The leader of the Party who escort the bride to the residence of the bridegroom on the night is called *Lawichal.* The bridegroom pays to *Lawichal* in cash or in kind according to mutual agreement. There are some sections of Paite who do not follow this practice.

U Kaan Man

When there are two or more girls in the same family, if the younger sister gets married before her elder sister, it is called *U Kaan* (that means, jumping over her elder sister). The bridegroom of the younger sister has to pay in cash or gift a shawl to the elder sister called *Angpuan* (Fig. 3.1) and one showl to the mother of the bride.[3]

Thaman

Thaman[4] is a small sum of Rs. 2 paid to the parents of the bride. After payment of this, the bride groom side has the right over the body of the bride and can bury her, in case she dies, unless this amount is paid, her parents may come and claim her body. Thaman is basically a symbolic sum, carried over from time immemorial.

Mouman (Bride Price)

It is paid in kind—three mithuns of which one is female and its calf must be included. Earlier, the custom was, whatever the Bride price was decided between both the parties, the bride price preferably in coins would be kept in *lawhpi* and brought to the bride's family, but now-a-days, the bridegroom may pay in paper money, *i.e.* in lieu of the 3 mithuns, the amount will be decided by the bride's family (Fig. 3.2).

Mou Vanken (What the Bride Carries with Her)

It was indispensable on the part of a bride in earlier times to bring with her the following articles to her husband's house:

(a) *Puanpi* (locally woven cotton blanket).
(b) A bottle containing edible oil called *Sa Thou Bing.*
(c) Tools and Implements like hoe and axe.
(d) A carrying basket with carrying strap made of cane.

Along with the above mentioned things, the bride may bring along with her things according to her wish.

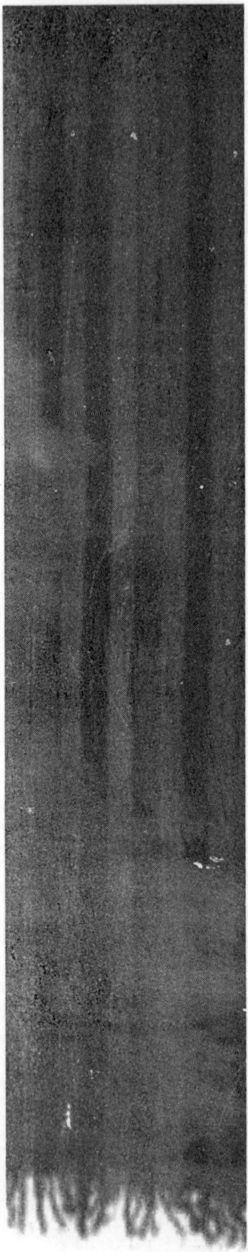

Fig. 3.1: Shawl (Angpuan)

Fig. 3.2: Earlier the Bride-Price was paid in Coins as shown above

Sial Khumsa

When the bride price has been paid, the parents of the bride are obliged to kill a domestic cattle, in honour of their daughter and invite the Household council of the bridegroom who, too, invites their relatives. This is called *Sial Khumsa Gou/Goh.*

Sisan Pallou

A married woman who has not yet given birth to a child is referred to as *Sisan Pallou* which when translated means "not yet experience bleeding at child birth".[5]

Divorce (Maak)

A divorce[6] between a husband and wife is neither sanctioned nor allowed by custom, but divorce can take place because:

(a) of adultery on the part of either the husband or wife.
(b) due to the wife not able to give birth.
(c) due to unsoundness of body and mind.

The amount to be paid for divorce will be decided by the two parties of the wife and husband after an agreement is reached.[7]

Remarriage of Divorced Woman

If and when a divorced woman re-marries[8] another man, her price should not exceed one mithun. If and when a married brother dies without leaving any issue and the brother of the dead man marries his wife, one mithun and *Puandum (a Paite shawl)* have to be given to the parents of the bride in addition to the ones already paid at the time of her first marriage. This is called *Levirate* in sociological terms, it is not a rule but depends on the respective family.

Among the *Paites,* there does not exist any system of polyandry, *i.e.* a woman having more than one husband and polygamy, *i.e.* a man having more than one wife, but the mule of *Monogamy, i.e.* a man having one wife, is followed.

Elopement (Zawltai)

If a boy[9] elopes with a girl but does not like to marry her afterwards, he is liable to a fine of one mithun with a *pot of rice beer (now, tea).*

Zawllei

If a man has a child with a woman but refuses to marry her, he is to be punished with a fine of three mithuns called *Zawllei.*

Tanu Sagoh

The parents of a bride may organise a feast in favour of a married daughter anytime, this is called *Tanu Sagoh.*

Family and Kinship

A man when he marries, if he is the eldest son he stays with his parents and look after his brothers and sisters, the younger brother usually moves out and establish a separate house when he gets married. But now-a-days, with a lot of migration taking place and occupational mobility, son do not always stay with their parents, they set up a nuclear home with their wives and children and the parents are looked after by relatives or they go to stay with their children.

Indongta (Household-Council)

It is essential for a married couple after having a separate house to have a Household Council. The Household Council takes up matters like death, loss, marriage etc.

The Household Council should consist of the following members:

(i) *Intek*—The *Householder.*
(ii) *Thallough*—A closely related brother who may be dear to the householder is appointed as *Thallouh.*
(iii) *Thusapi*—A reliable person from other clan is generally appointed as *Thusapi* (first), others *thusa* may be appointed as and when necessary like:
Thusa Nihna (Second)
Thusa Thumna (Third)
(iv) *Tanupi*—Householder's elder sister is appointed as *Tanupi.*
Tanu Nihna—Householder's second sister, if any,
Tanu Thumna—Householder's third sister, if any.
(v) *Pupi*—Mother's brother of householder is called *Pupi.*
(vi) *Pu Nau*—The wife's brother or in-laws of householder is called *Punau* or *Punou.*
(vii) *Behval*—The 4th *Thusa* having no specific assignment in the distribution of works is called *Behval.*
(viii) *Zawl*—A friend of householder but of other clan is appointed as *Zawl.*

Tou Sagoh

It is a feast arranged by a married younger brother by killing a four-footed animal to the house of the eldest brother as a token of respect to his elder brother and as a symbol of making an independent household on the part of the younger brother. When such feast is arranged, the eldest brother gives certain properties to the younger brother to start an independent life. If a younger brother does not arrange such a feast then it is presumed that it is disrespect towards his eldest brother. But no fine is imposed for not performing such *Tousa.*

Puzukholh

A nephew pays respect to his maternal uncle by bringing *Zu* (rice beer but now tea) as a token of love and respect, the latter gives blessing to his nephew and organises a feast in honour of his nephew.

When a child is born to a married couple, if it is a boy it is given name by the child's grandfather or any male member of the child's father. The naming is done by taking the last name of the person who has named the child.

For example, if the name of the child's grandfather is *Thangkhanpau,* the child's name will start from Pau, if the child is a girl, likewise, the grandmother of the child or any female member of the father or mother's side will give name to the child.

Pudem/Simmoh (Challenging an Uncle)

If a nephew *(Tupa)* refuses to give a due share of four-footed domestic or wild animal to his maternal uncle (Pu) he is said to be *Pudem/Simmoh* and he is liable to pay a fine of upto one Mithun.[10]

If a man deserts his wife and children and the fact is confirmed by authority, all his properties will be made over to his wife and children and he may not claim the price of his daughter because he has not looked after them.[11]

When a woman is married she is expected to leave behind all her customs and follow in the footsteps of her husband. The *Paite* tribe follows a *Patrilineal Descent Line.* The male is the head of the family, a woman moves out at the time of her marriage, and children born belong to the husband's family.

The *Paite* tribe is a *Patrilineal* society. The succession is through the male line. The eldest son is entitled to succeed to the properties of his father at the time of his death; but the father has got a right to divide/distribute his properties to other sons before he dies.

In case, the Head of the family dies without making distribution of his properties to his children, then the Household Council in the presence of the children can make distribution of the properties to the children of the deceased father.

Gou

A property[12] both acquired and inherited by the Head of family, movable and immovable which remains at the time of his death is to be inherited by his own son or sons, this property is called *Gou.*

Gam

If the head of family has no son of his own at the time of his death to succeed or inherit his properties both movable and immovable, then the properties are known as *Gam* which will go to the nearest male member of his relative.

Gamluah

If the Head of the family has no son of his own living at the time of his death, then his properties will go to his next brother, and if he has no brother; then it shall go to the nearest male member of his relative.[13]

The *Paite* tribe also have features of what *Morgan* wrote about *Classificatory and Descriptive Systems of Kinship.*

When *Morgan* talked about *Classificatory system,* several people, lineal as well as collateral and often, even affinal are all referred to the same term of designation, *e.g.* the sisters of the brother among the Paite tribe is referred to as *'Ni'* and the husbands of the sisters are referred to as *'Gang'* (Fig. 3.3).

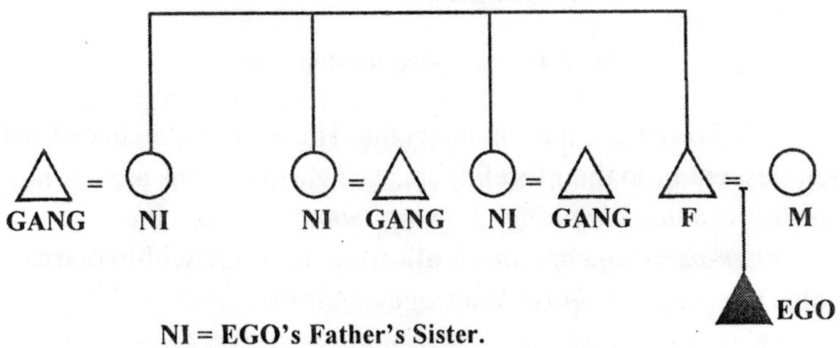

NI = EGO's Father's Sister.
GANG = EGO's Father's Sister's Husband

Fig. 3.3: Classificatory Kinship Terms

Descriptive System of Kinship refers to the describing of the person relation who is addressed, *e.g.* among the *Paite* tribe, the father's eldest brother is referred to as *'Papi'* and his younger brother *'Paneu'* the wife of *'Papi'* is referred to as *'Nupi'* and the wife of *'Paneu'* is referred to as *'Nuneu'* (Fig. 3.4).

The Paite tribe also follows a practice of what is called *Patrilateral Cross Cousin's Marriage, Paite* refer to it as *Ni Nung Zui,* which when translated means following the father's sister (Fig. 3.5).

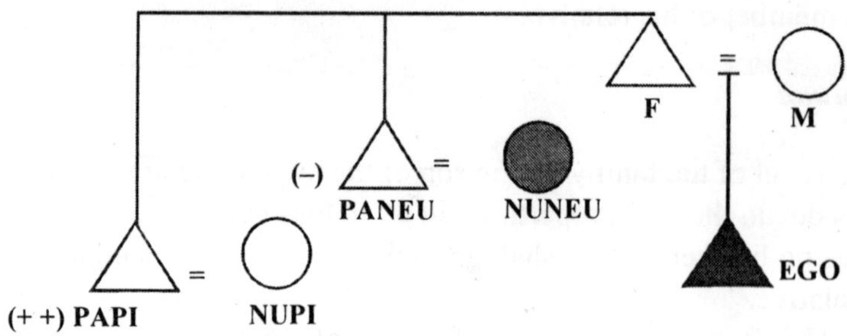

(+ +) PAPI = EGO's Father's Eldest Brother.

NUPI = PAPI's Wife.

(–) PANEU = EGO's Father's Younger Brother.

NUNEU = PANEU's Wife.

F = EGO's Father.

M = EGO's Mother

Fig. 3.4: Descriptive Kinship Terms

The *Paite* tribe have many clans. These clans can each trace their descent upto their forefathers. The various clans are—*Guite, Ngaihte, Samte, Khuptong, Tonsing, Shoute, Phaipi etc.*

Inter-marriage of clans is allowed, marriage within one clan is also allowed, *i.e. Intra-Marriage of Clans.*

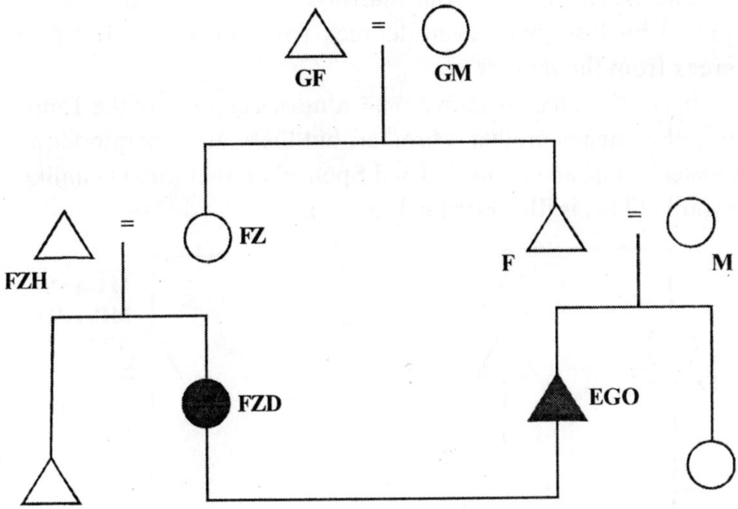

Fig. 3.5: Patrilateral Cross Cousin Marriage

Paite Kinship

It would help us to know more clearly about the Paite's Kinship network, if we examine the North Indian kinship system and the South Indian kinship system and see which system of kinship the Paite fall into.

In the South Indian rules of kinship[14] they can marry their cross-cousins—a class of kin which includes the Mother's Brother's child and the Father's Sister's child and, at the same time, are forbidden to marry parallel kin, including brother and sister, father's brother's child and mother's sister's child. In South India, especially among the Tamil, Māman is Mother's Brother, Father's Sister's Husband and Spouse' Father (Fig. 3.6).

In North India, it is prohibited degrees of kind of relationship and not proxity which governs the rules of marriage. Cross-cousin marriage is forbidden in the North, and it is extremely likely that people in the North marry at a greater distance, spatially and structurally, than in the South.

The North Indian avoid marriage with somebody who is removed by less than seven degrees from the father and five degrees from the mother.

In North India, the term *Māmā,* almost a cognate of the Tamil word, also means mother's brother, but there are seperate terms for sister's husband (*Phupha*) and Spouse's father (*Sasur*) unlike the south. (This is illustrated in Fig. 3.7).

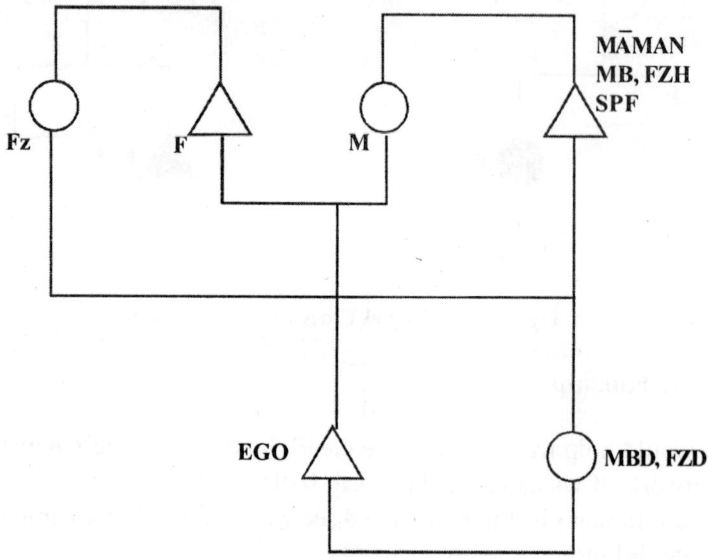

Fig. 3.6: South

In North India, the members of the Ego's generation, cousins of all kinds are regarded as so many species of *Bhāi* (Brothers) and *Bahin* (Sisters), terms which designate one's own brother and sister, cousins are not, as in South India, divided into two large categories of cross and parallel, and relatives by marriage are, on the whole, kept distinct from relatives by blood.

In the North India, it is the parents of the Groom and Bride who arrange the marriage. The North has separate words for *'Daughters'* and *'Brides'* in each regional language. The custom of local exogamy divides the women of a local group into two sharp divisions: The *'Daughters'* of the village and the *'Brides'* of the village. The Bride has to almost always have to contend with her *'Sās'* (Mother-in-law), and *Nanad* (Husband's Sister).

In the South, the distinction between *'Daughters'* and *'Brides'* or *'Wives'* is not as deep as in the North. A Girl's behaviour in her husband's family is much freer. After all, her husband's mother is either her own Grandmother or her Aunt.[15]

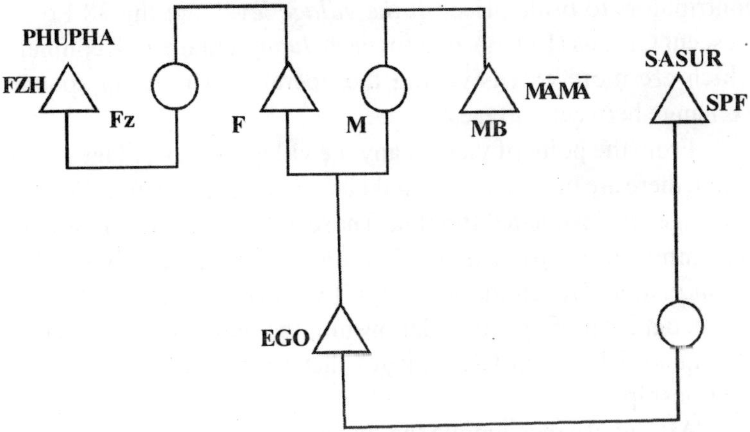

Fig. 3.7: North

The Paite kinship network to a limited extent is quite similar to the South Indian kinship network.

Among the Paites, there is an existence of Cross-Cousin's marriage. Like the South, kinship system the Bride is relatively free at her in-laws home because most of the time, marriage is among families known to each other.

There are tribes who do not fall into the North or South kinship system.

Likewise, the so-called 'Tribals'[16] of Middle India do not fall nearly into the usual division between South India and North India kinship network. This division has itself long been recognised and is, *inter alia,* one between areas that value and do not value Cross-Cousin as a mode of affinal alliance but the tribal system of Madhya Pradesh, Southern Bihar and Orissa also have a number of distinctive features, so much, so that, at first sight they suggest the existence of a third kinship model for the subcontinent.

In *Keonjhar,* the *Juang* a tribe of Northern Orissa, there are two significant levels of segmentation. *At the lower level,* are the unnamed *Kutumali* (an Oriya word), which consists of four or five agnatically related families and are important at Life-Crisis Rites, as the preferential unit of inheritance and as the main collective contributors to bride price. *At the village level,* are the 38 Local Descent Groups (LDGs), one for each *Juang* village in Keonjhar, which are more extensive and the operational units in spouse exchange between villages.

From the point of view of anyone village, other villages with whom there are or potentially could be relations of alliance through marriage are designated Bondhu. Those with whom such relations are impossible, are called *Kutumb,* which means basically *'Unmarriageable'* in *Juang,* but in two different senses: One is the recognition of agnatic relationship; the other is the existence of a quarrel between two villages that were once in a *Bondhu* relationships.

Affinal alliance among the *Juang,* described by McDougal, as involving *Classificatory Sister Exchange.*

Affinal alliance among the *Juang,* should only take place between members of LDGs and villages that are *Bondhu* to each other and members of the same *Bhaiguli.* Marriage with anyone in another *Kutumb* group would be *Bogudung* or *Incest*: Marriage with anyone in the opposite *Bhaiguli* would be *Jenkani,* literally *'Adultery'*.[17]

NOTES

1. The Paite Customary Laws (1986).
2. The Paite Customary Laws, *op. cit.* no. 1, pp. 15-18.
3. *Ibid.*
4. G. Piangzathang, "The Story of the Paite in Indo-Myanmar" (unpublished).
5. *Ibid.*
6. The Paite Customary Laws, *op. cit.* no. 1, p. 20.
7. *Ibid.*

8. The Paite Customary Laws, *op. cit.* no. 1, pp. 5-8.
9. *Ibid.*
10. *Ibid.*
11. G. Piangzathang, *op. cit.* no. 2, p. 77.
12. The Paite Customary Laws, *op. cit.* no. 1, p. 26.
13. *Ibid.*
14. Patricia Oberoi, *Family, Kinship and Marriage in India,* (1993), Edited.
15. *Ibid.*
16. Contributions to Indian Sociology (N.S.) 24, 1 (1990), pp. 61-76.
17. *Ibid.*

4

Political Organization and Social Control

The kinship network of the *Paite* tribe is spread across the village. The village is the main epicentre around which various functions and duties are undertaken.

Hausa (Chief)

He was the founder[1] and head of a village and ownership of village land belongs to him (Fig. 4.1)

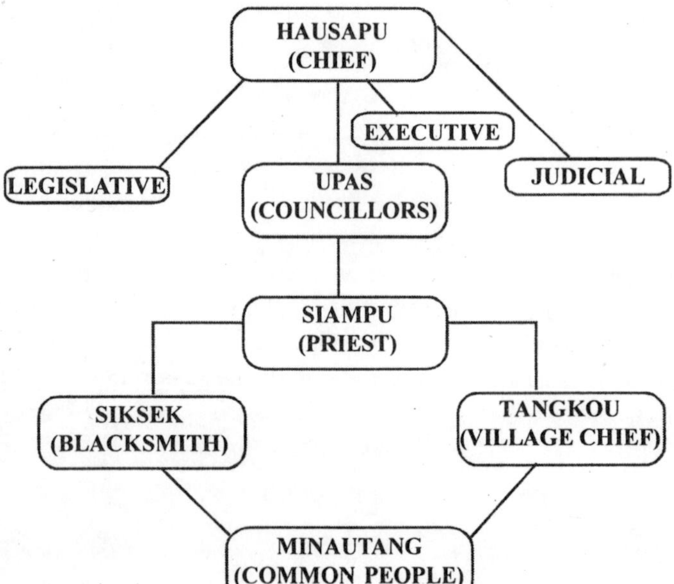

Fig. 4.1: Village Structure

We will discuss the Paite traditional set-up of Political Organisation.

Powers of Chief

The chief has sovereign powers—legislative, executive as well as judiciary in his village. From minor case to criminal cases like murder etc. are decided by him with the help of his councillors called UPAS.

Hausa Sutpi Kawi

Criminals who are afraid of being killed by others run into the house of a chief and embrace a supporting host inside the main room. After, the criminal has entered the chief's house and embraced a supporting host no man can hurt him without proper trial of his case by the chief. Action has to be undertaken according to the decision of the chief.[2]

Executive Powers

The chief was the executive head[3] of the village. He was sovereign in the political affairs of the village. His decision on all important matters used to be final. He was the appointing authority for all important posts of the village. He used to appoint the village priest, the village blacksmith, the members of the village council locally known as the *UPAS* to assist him in the village administration.

The chief was properly looked after and maintained by the villagers. He used to act as the guardian and administrator of the village. The people looked upon him as the protector of their lives and properties. His position in the village was unique in the sense that he enjoy absolute powers.

Legislative Powers

The chief enjoyed enormous legislative powers, all matters relating to issuing of ordinances, framing of rules and regulations and important instructions were made by the chief. The rules and regulations were based upon the customs and traditions of the people. As such, these, ordinances were strictly observed and carried out by the villagers.

The main duty of a villager was to obey the orders of the chief. Those who disobeyed were severely punished or expelled from the village. The chief had the power to award death sentences to those who had committed great crimes. Though the chief had the power to give death sentences, yet he never did so and instead he preferred to expel the offender from the village.

The chief used to issue orders every year regarding the selection and distribution of *Jhum** land to the villagers. On the chief's orders, the villages used to construct and maintain the village water tanks.

He used to fix the time and date for observance of traditional festivals. In times of drought and famines, the chief was empowered to issue orders and instruct the villagers regarding economising consumption of water or stoppage of rice export to other villages.

Judicial Powers

The chief was the supreme judge in deciding all civil and criminal cases of the village. The chief assumed two important judicial roles—first, as interpreter of the customary laws of the people, and secondly, as judge within its respective chiefdom.

The village court of justice was composed of the chief and his council members *(UPAS)*. The village court held its meetings in the chief's house and was presided over by the chief himself. The proceedings of the cases tried by the village court were never put down in writing as the chief and his *UPAS* were usually illiterate.

Cases relating the theft, murder, adultery etc. were tried by the chief. His decision was considered final. As far as cases or disputes relating to neighbouring villages were concerned the chiefs of the two villages used to sit together and take decisions. The amount of fine imposed varied from case to case.

* Zhum—Cultivable Land.

Military Powers

The *Chief* used to be the *Commander-in-Chief* of the village army. The success or failure in war with other tribes or villages depended entirely on the chief's capability.

He could make an alliance with the chief of another village in order to counter the village chief who used to work against his interests.

In the village, *a dormitory* (locally known as *Haam*) was established in the *Chief's* house. The dormitory used to be a common lodging place for all the young and unmarried males. Usually, those who stayed in the dormitory constituted the chief's village force.

The dormitory also had other two divisions other than the one occupied by the youth. The other two divisions were:

(i) **The Haam UPA (Adult Members):** They acted upon the orders of the chief.

(ii) **Haam occupied by Youngsters (Naupangte):** They were duty bound to carry out the orders of the *Youth (Tangvalte)*. They used to wash shirts and pants of the youth members of the dormitory. They collected fire woods and prepared the beds of the *Youth* and *Adults*.

The *Chief* had the priviledge of receiving various taxes from the villagers like:

(i) Paddy Tax (Buhsun)

It was the duty of every household in the village to pay at least a tax of one basket of rice per annum for use by the chief's household, the tax usually varied.

(ii) Meat Tax (Saliang)

Whenever, a villager kills a four-footed wild animal, a portion of the meat was paid to the chief as tax.

(iii) Salt Tax (Chi Siah)

Whenever salt was found in the village, the chief was entitled to a substantial share of all the collections made therefrom.

(iv) Bee Tax (Khuai Siah)

If a villager extracted honey, the chief was paid his share, being the owner of the land.

Limitations and Checks on the Chief's Power

Although the chief enjoyed enormous powers, these are not without limitations. If the chief became tyrannical or indifferent to the welfare of his subjects or inefficient in tackling the administration, the villagers could desert him at any time.

The usual way by which the chief could be checked was through the mass migration of the villagers to another village. But such migration of the villagers to another village without the prior permission of the chief would empower the latter to confiscate the villagers movable and immovable properties.

Moreover, if a villager was expelled from the village by the chief due to certain crimes commited by him the chief was empowered to confiscate all the properties of that villager.

Even though the villagers were not satisfied with the chief's descision yet they had no right to oppose the chief individually or collectively. So, the only alternative is to migrate to another village. The village chief had no right to stop the villagers' migration to another village as custom sanctioned such a practice.

The powers of the chief could also be curtailed by the village council or UPAS. If the UPAS are intelligent and capable of representing the chief in administration they could make the latter to act just like a puppet in their hands. In this way, the UPAS could ultimately act as a factor in curtailing the chief's powers.

In addition to this, when a chief was defeated by another chief, the latter used to control the defeated chief in his administration. The victorious chief used to demand different kinds of taxes from the defeated chief which were to be paid annually.

A small village chief simply survived at the mercy of a bigger and more powerful village chief. A skilful chief enjoyed a feeling of superiority in his relations with the various neighbouring chiefs.[4]

UPAS of Chief (Councillors)

The chief of the village[5] can nominate anyone of his villagers who he thinks fit to be his UPAS. The UPAS are to assist him in discharging his duties and at the same time they are to execute the chief's orders.

The UPAS have the priviledge of getting a section of Jhum lands and exemption from labour. Each village has a *priest* or a *sorcerer.* The post of a priest is not hereditary.

His duty is to look after the sick and make sacrifices to the gods for their recovery. Each village also has at least one blacksmith who makes and repairs hoes, spades, axes and other implements and tools.

A village has a man called *Tangkou (Village crier)* who cries to the villagers for the information of the chief's orders and other notices. He walks around the village crying aloud to the hearing of the villagers.

Distribution of Jhum Land

Village land belongs to the chief. Villagers can have houses and jhum on it. *Chief,* with the help of his *UPAS,* set apart every year certain portion of the land for jhuming for the year.

Chief, his *UPAS* and village crier, just select jhum lands and then the general public follow suit. Jhum cultivation is also called slash and burn cultivation.[5]

Let us study the *Gonds.* I took the *Gonds* to see the differences or similarities, if any, between a *North-Eastern Tribe* and a *Non-North Eastern Tribe.*

Gonds appear not only[6] as the main population of wide areas in the very heart of India, described after them as *Gondwana.* Spread over an area considerably larger than the British Isles and extending from the *Godavari Gorges* in the South to the *Vindhya*

Mountains in the *North,* the *Gonds* are neither racially, nor culturally, nor linguistically a homogenous population. Wherever we find *Gonds,* unless they are totally detribalized and merged with *Hindu* population, they describe themselves as *Gonds* or, if speaking *Gondi,* as *Koitur,* the universal equivalent of that name in all *Gondi* dialects.

I will compare the *Gonds* and the *Paites.* The *Gonds* are one of the *largest Tribe* in India whereas the *Paites* are a *comparatively smaller Tribe.*

The *Gonds* written about here are of the *Adilabad* district in *Andhra Pradesh.*

The village is a social unit for religious and social purposes. The village community consisted of an aggregation of elementary families, membership of a village community carries with it not only rights, but also obligations, there are negative and positive obligations, there are unifying devices promoting and cooperation in the cause of village welfare.

There are *Two* dignitaries who are the symbols of unity of the community. They are:

The Village Headman (Patla): Who *act* as its *Representative* and *Leader* in *Secular Affairs.* He maintains a *Good Spirit* of *Amity* and *Co-operation* among the *villagers.* He *settles quarrel* among the villagers. He has the *Right* to *One Day's Free Labour* on the *Part of all villagers.*

The Village Priest (Devari): The *Devari* is the one, responsible for *Ritual Acts* and he is the *Spokesman* and *Mediator* between the *Community* and the *Dieties* of the *locality.* He is *exempted* from the contribution of *Sacrificial Animals* and *Grains* for the *communal rites.*

The Village Council or Village Panch: The village *Panch* is the *Embodiment* of *Authority* and stands in a certain sense for *Gond* society as a whole. The *principle functions of the Panch,* is the *Preservation of the Social Equilibrium.* It is for the *Panch* to give sanction for the celebration of a wedding or the performance of a *Karun,* which is the *Final Funerary Rite.*

The invariable and permanent elements of the social structure of the *Gonds* is the division into *Four Exogamous Patrilineal* descent groups known as *Saga.* The *Four Phratry* system is traced back to the dawn of *Gond* history.

For a time the people of *Four Saga* dwelt together but as the population increased, the harmony between the intermarrying groups was broken by disputes over matters of *Precedence Leadership*, so the *Four Phratries* decided to settle in separate villages.

Among the *Gonds* men of a Clan (*PARI*) consider themselves as agnatically linked kinsmen and use the appropriate kinship terms. Clans are sub-divided into *Sub-Clans* or *Khandan.* Every *Clan* is strictly *Exogamous.* Clan incest is never heard of.

Marriage

Every man and woman, rich or poor, should marry with full rites, known as *Marming,* once in his life. The *Gond* does not reach the full status of a member until he has undergone the *Marming Rites.*

There are *Three Legitimate* ways of getting married, they are:

(a) Marriage by Negotiation (*Khaja Khopra Marming*)
(b) Marriage by Service (*Lamsare Marming*)
(c) Marriage by Capture (*Pisiwatwal Marming*)

Marriage by *Negotiation* has the greatest prestige value, it involves a formal proposal, the subsequent *Betrothal ceremony* (*Khaja*) an *Elaborate Wedding* (*Marming*) in the bridegroom's village with the kinsfolk of both marriage partners participating, and ideally also the immediate payment of *Bride Price* (*Pari Kapra*).

A family unable or unwilling to pay the bride price immediately will attempt to obtain a bride by capture only a man confident of his ability to bear the expenditure on the betrothal ceremony. The wedding and the bride price will formally ask for a girl to be given to his son in marriage.

The preliminary proposal is followed by a solemn betrothal ceremony, known as *Khaja,* for which a large party from the

Groom's village journeys to the house of the Bride's parents. There the bride price is agreed upon, and a day for the wedding is fixed.

As *Gond* society is basically patrilineal, the majority of young married couples live for some time with the husband's parents.

Divorce is not infrequent and the *Gonds* do not have much difficulty in remarrying.

The rules of inheritance are clearly those of a patrilineal family system with little regard for the interests of women. In principle, property is handed down from father to son. If there are sons in a family the bulk of the property is divided among them while the daughters may get only some insignificant possession.

Kinship

The kinship system of the *Gonds* consists of isolating, descriptive and classificatory terms, the latter constituting the majority of all kinship terms. Isolating terms, which designate one person only, are confined to the words for husband and wife, both applicable only to a person's actual mates and not to any of the latter's relatives. All other terms are classificatory and cover several persons standing to the speaker in different biological relations.

On the whole the *Gond* kinship system is consistent within itself, and the only feature which calls for explanation is the use of the identical term for certain affinal relatives separated by two generations. Ego refers to his wife's elder brother as *Tad Murial,* and uses the same term to denote his wife's father.

Though Phratry and Clan Exogamy are primarily based on the principle of Patrilineal descent, there is also a kind of Exogamy of the female lines. A man may not marry a girl whose mother belonged to the same (*Pari*) as his own mother.

Religious Beliefs and Practices

The *Gonds* have many deities, the main is called by the term *Persa Pen* or *Great God.* Among many deities is the village mother called by the term *Nat Auwal.*[6]

NOTES

1. G. Piangzathang, *The Story of the Paite in Indo-Myanmar* (Unpublished), 1995.
2. *Ibid.*
3. *Ibid.*
4. Chinkholian Guite, *Political-Economic Development of the Tribals of Manipur,* (1999).
5. G. Piangzathang, *op. cit.* no. 1, pp. 69-71.
6. Christoph von Furer-Haimendorf, *Gonds of Andhra Pradesh,* (1979).

5

Religion and Social Change

Religion serves the central and crucial function in society, supporting what has been variously called social integration, social solidarity and social cohesion.

Religion is an all-pervading supernatural phenomenon in man's life. For this reason perhaps it has been considered universal. Religion has exercised the most profound influence over man's thoughts and behaviour from times beyond human recollection.

NUER Religion by Evans-Pritchard (1956)

Religion was explained in terms of association of ideas, of personification of natural phenomena, of awe, of thrill, of fear, anxiety and frustration and so forth.

Nuer religion indicates a distinctive kind of diety which is dominated by a strong sense of dependence on God and confidence in him rather than in any human powers and endevours.

Among the *Paite* tribe, religion in the form of Christianity has played a significant role in bringing about change and development. The change that has been brought about, can be trace to being somewhat similar to what *Srinivas* termed as *Westernisation.*

Westernisation refers to the changes brought about in Indian society and culture as a result of over 150 years of British rule, the term subsuming changes occurring at different levels, *i.e.* technology, institution, ideology and values.

Christianity has brought about changes in the family, marriage, social life of the Paites, their lifestyles, food habits has also undergone significant changes.

Christianity did not exist before, it was the *British* missionaries who took up the work of converting the *Paite* tribal people from their previous beliefs in animism and the supernaturalism.

Durkheim defines religion as, "A religion is a unified system of beliefs and practices relative to sacred things that is to say, things set apart and forbidden—beliefs and practices which unite into one single community called a church, all those who adheres to it".[1]

Johnstone observed that religion[2] is a "System of beliefs and practices by which a group of people interprets and responds to what they feel is supernatural and sacred".

In and for the conduct of his daily life man has evolved certain mechanisms. Thus, there are his economic organisation and technology. Then, there is religion—it is the human response to the apprehension of something or power, which is supernatural and suprasensory.[3]

Radcliffe-Brown takes a different stand by saying the "Function of religion is not to purge fear and other emotional strains from the human mind but to instil a sense of dependence in it."

Marx observed "Religion is the general theory of that world, its encyclopaedic compendium, its logic in popular form, its spiritualistic point, however, its enthusiasm, its moral sanction, its solemn complement, its universal for consolation and justification. It is the fantastic realization of the struggle against religion is, therefore, immediately the fight against the other world, of which religion is the spiritual aroma. Then, follow the well-known and often-quoted life. Religion distress is at the same time the expression of real duties and a protest against real distress. "Religion is the sigh of the opressed creature, the sentiment of a heartless world, just as it is the spirit of spiritless situations. It is the opinion of the people".[4]

Religion is expressed in two ways: by the ideas and convictions about God or the Supreme Super-natural Being forming a base philosophy and by a system of prayers, rituals and obligations in the name of the God, his colleagues and his hierarchical subordinates culminating in some kind of elaborate ceremonial.

Religion is a subject of great concern even in advanced society of today. Supernatural beliefs are present in every known society. Their variety may be endless.

Emile Durkheim argues[5] that all societies divide the world into two categories 'the sacred' and the 'profane' or more simply, the sacred and non-sacred. *Religion* is based upon this division. It is important to realize that, 'By sacred-things one must not understand simply those personal things which are called gods or spirits; a rock, a tree, a spring, a pebble, a piece of wood, a house, in a world anything can be sacred.

There is nothing about the particular qualities of a pebble or a tree which makes them sacred. Therefore sacred things must be symbols, they represent something. To understand the role of religion in society, the relationship between sacred symbols and that which they represent must be established.[6]

Radcliffe Brown emphasised[7] that ultimately, the survival of the group is more important than that of the individual; and if the latter has to make some sacrifices in his own interest to do so, because without social survival, individual survival is not possible.

However, the individual does not seem to realize this always and he seeks to chart out an individual course of action. If each individual were to do this there will be better confusion and chaos and no organized activity would be possible.

Adherence to a norm of behaviour is essential in terms of social survival; and it is the fear of supernatural control and punishment, as also the anticipation of support in the case of socially approved conduct, that brings about their adherence.

Therefore, the function of religion is to create a two-fold feeling of dependence on society and thereby obtain the individual's concurrence with the social norms, the ultimate aim being social survival. The function of religion is the contribution it makes to that total activity which is designed to perpetuate society.[8]

Weber is largely concerned with a single major aspect of religious ethics, namely their connection with the economic order. The part played by Calvinist ethics in the development of modern capitalism. The Calvinist is never certain that he is one of the elected persons and looks for signs of his election in his earthly life and he finds them in the prosperity of his enterprise.

But he is not permitted to enjoy leisure as a result of his success or to use his money in the pursuit of luxury or pleasure. Thus, he is obliged to re-employ his money in his business and the formation of capital takes place as a result of this ascetic obligation to save.

"Secularization" is now-a-days generally employed to refer to in the word of Peter Berger, 1973:113 "the process by which sectors of society and culture are removed from the domination of religious institutions and symbols".

The word *"Secularization"* was first used in 1948, at the end of 30 years of War in Europe to refer to the transfer of church properties to the exclusive control of the Princes.[9]

Being modern means larger and deeper things; for example, the enlargement of human freedom and the enhancement of the range of choices open to people in respect to things that matter including their present and future lifestyles. This means change of one self. This is one of the connotations of secularization.

Pre-Christian Period

The hill people of the North East from the distant past up to the time of their conversion to Christianity had a *certain concept of religion.* They believed in some spirit or powerful being which had its abode in natural objects like stone, river, tree, mountain peak etc. There were many spirits benevolent as well as evil which influenced the life of man.

These spirits had to be kept happy by sacrifice. In every village, there were priests who were conversant with process of offering sacrifices. The hill men also worshipped a Supreme God who enjoyed the highest position. Next to God there were quite a large number of spirits and demons which deserved to be worshipped.

These spirits were local demigods and were thought to reside or haunt particular localities: village, river, stone, forest, mountain tops and so on and so forth. In one word, the tribal people may aptly be called animists. They also worshipped the ancestors and offered sacrifices to please the souls of the dead.

Heaven to them is a place for righteous people. The spirits are propitiated for guarding them against evil influences, dangers to their life, to cure sickness etc. Important sacrifices were connected for the prosperity of the village, for good yield of crops, to cure barrenness of women. They believed in black magic.

Across the golden heights[10] of Assam and NEFA*, most of the Eastern tribal people have their own tribal god and goddesses and have deep respect for other faiths. Assam and the neighbouring areas have been a home of various forms of religious beliefs since time immemorial, while the plains have always been under the influence of cults prevalent in other parts of India, the ribal areas, exept in a few cases have remained free from them.

Nearly, all the tribal people believe in the existence of spirits that move nature and guide human actions. They give different names to them and worship them in their own distinct way. The priests occupy a very important position in the tribal society. No ceremonies are performed without them. Annual sacrifices are popular with the people in the hills. According to their beliefs, gods have to be worshipped with their favourite animals.[11]

Among the tribal people, religion becomes all the more important for it is inter-woven in their entire social life and shapes most of their social behaviour. The supernatural becomes the motivator and custodian of all their deeds, tries to discover and rule over many new places of human dwelling and inspires nobel thoughts and noble actions.

Malinowski, who worked with the primitive *Trobriand Islanders,* poitns out that with reference to the Trobriand Islanders, religion is ultimately connected with various emotional states, which are states of tensions. For example, quite a few of their magical and religious practices centre around fishing expeditions. These are the outcome of the state of fear which a possible disaster on the seas give rise to.

Similarly, hate, greed, anger, love etc. may arise due to various situations in a man's life. These situations created stress and strains.

---

* NEFA—North-East Frontier Agency, of which Manipur was a part.

Religion is made use of in such a situation as a tool of adaptation, its purpose is to purge the human mind of its stress and strain, *i.e.* it is cathartic in its action. In other words, religion has the function, of bringing about a readjustment between man and the supernatural upset states of existence.

Definition of Social Change

Social change is the significant[12] alteration of social structures (that is, of patterns of social action and interaction) including consequences and manifestations of such structures embodied in norms (rules of conduct) values and cultural products and symbols.

Social Change is such a prevalent and often disturbing feature of contemporary life that both the specialist and the layman may be tempted to suppose that it is peculiarly modern. Certainly the extent and rate of change in the modern world are greater than in most past periods. Change at some level and degree is a characteristic of man's life.[13]

In their book, *Character and Social Structure, Gerth* and *Mills*[14] have outlined *six* major questions which can be asked about social changes:

(i) What is that changes?
(ii) How does it change?
(iii) What is the direction of change?
(iv) What is the rate of change?
(v) Why did change occur or why was it possible?
(vi) What are the principal factors in social change?

In dealing with the first of these questions it is useful to define social change, as a change in social structure or in particular social institutions or in the relationships between the institutions.

The question concerning the manner, direction and rate of change require for their answer historical description and interpretation, such as have been provided, for example in the various accounts of population changes of the increasing dimension of labour in industrial societies, of change in the character of the modern western family and so on.

Gins Berg has undertaken a systematic analysis of the factors which have been invoked by the different writers to explain social change.

The conscious desires and decisions of individuals (exemplified by the development of the small family system in western countries). Individual acts influenced by changing conditions (*e.g.* the decline of villeinage in England between 1300 and 1500 A.D.).

Structural changes and structural strains. [contradictions between forces of production and relations of production emphasized by Marxists].

External influence (cultural contact or conquest). Outstanding individuals or groups of individuals a confluence or allocation of elements from different sources converging at a given point, *e.g.* revolution.

Fortuitous occurrences *e.g.* [the Norman's conquest of England, the British conquest of India] the emergence of a common purpose.[15]

The *Paites* are undergoing a process of modernization. Mental attitudes and institutional structures constitute the key elements in the process of modernization.

Combining the concepts of innovation[16] and order, this mental attitude beings to develop once the modernization process has taken off. Dube identifies three connected and interactive aspects that are basic to the process. In his words, they are:

(i) A tested conviction of the existence of correlations and causes that maintains a continuing, systematic and inventive search for knowledge in other words, an analytico-causal and inventive outlook;

(ii) The multiplication of tools and techniques that both results from the first aspect and nourishes it; and

(iii) A willingness to accept continuous change on the plane of both individual and social structures, together with a capacity, meanwhile to preserve individual identity.

A series of social changes are implicit in the process of modernization. Modernity or Development is attained after the process of Modernization.[17]

Introduction of Christianity

Introduction of Christianity[18] resulted in certain changes in the structure of the tribal society. Prior to Christianity, the village Headman, functioned as the religious guide of his village, but after the advent of Christianity, the authority of church assumed higher importance in the religious matter. It inculcated unity among the tribesmen of the same origin. They started keeping Christian names.

Because of the contacts with the *British Baptist Christian Missionaries* there were various changes among the tribals. The missionary taught them carpentry, tailoring, printing, baking. Formal centres were opened by the missionaries who impacted academic and vocational training and introduced a system of education.[19]

Christianity in Manipur

Among the hill tribes of *Manipur,* Christianity has been one of the most significant in bringing about change and development. Majority of the hill tribes are Christians. This is true for the *Paite* tribe as well. The percentage of Christians in the hilly areas of *Manipur* is between 85 per cent to 95 per cent (Fig. 5.1 and Fig. 5.2).

Fig. 5.1: Manipur Christian Population

State/ District	*Total Population*	*Christian Population*	*Percentage of Christian Population*
Manipur	14,20,953	1,21,702	29.67
Manipur North	1,55,421	1,05,026	67.57
Manipur West	62,289	55,619	89.29
Manipur South (Mainly Paite)	1,34,776	1,18,887	88.21
Tengnoupal	56,444	44,718	79.22
Manipur Central9,	29,077	20,760	2.23
Imphal (M)	1,56,622	8,775	5.60
Manipur East	82,946	76,692	92.46

Note: (M) stands for Municipal Area.

Source: Registrar General and Census Commissioner of India, Census of India 1981, Paper 4 of 1984 (Delhi: Controller of Publications, 1984), pp. 446-448.

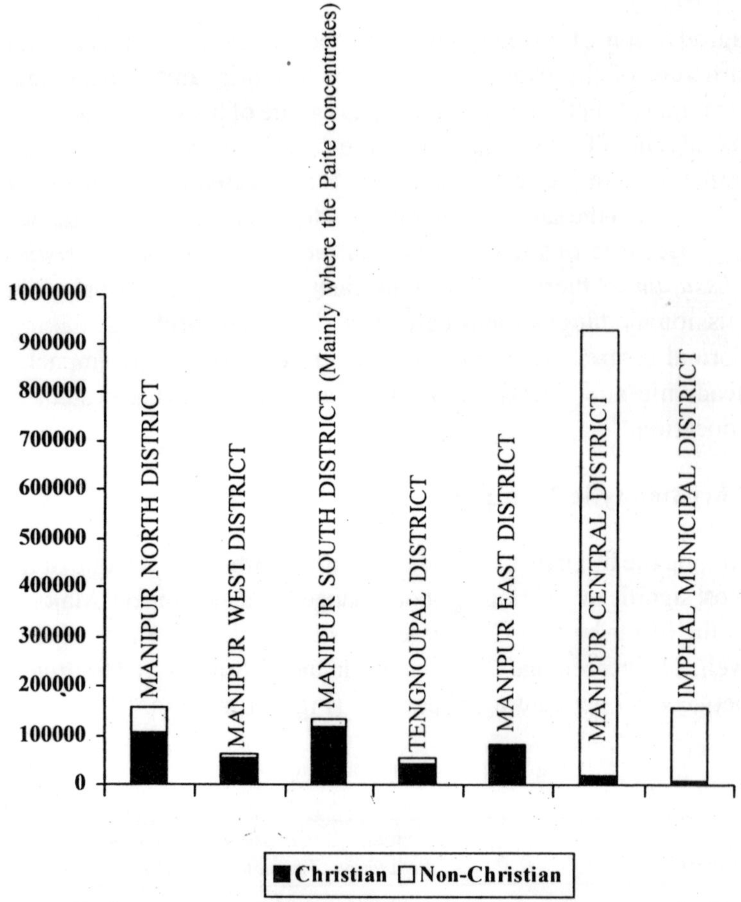

Fig. 5.1: Manipur Christian and Non-Christian Population 1981

How Christianity Started and Spread in Manipur

William Pettigrew was the first Christian[20] missionary in *Manipur*. He arrived at *Imphal* on *February 6, 1894*. He was not allowed to work in the capital city of *Imphal*, so he chose to work in *Ukhrul*, one of the district of *Manipur*, from this base the work was extended into the *Kuki** area and other neighbouring tribes.

* The Paite tribe was one of the tribe among Kuki (see p. 2 of Introduction).

He started his work by establishing a school in *Ukhrul.* The progress of conversion up to *1911* was slow. The converts were mainly young men reading in the school. The rate of converts picked by *1921,* so long, the converts included individuals but also groups and even whose villages came into the fold, the number of schools multiplied so did the converts.

The first *Kuki Baptist Church** was established in 1916. Besides people, the village chief also accepted the new faith. The people of neighbouring villages also became Christians. The Kuki chiefs exercised enormous power and if they accepted the new faith, the whole village would accept the same.

This paved the way for large scale conversion. The missionaries set up a middle English school, one orphanage, one leper/asylum and a dispensary. The missionaries learnt the tribal language and communicated with the tribals in their language/ dialect.

The North East India General Mission under Rev. Roberts pushed its work of conversion with full vigour. Churches were established among various tribes, namely *Paite, Hmar, Vaiphei, Kuki, Gangte*[21] etc.

The Work of Christian Missionaries in Manipur (Among the Paite Tribe)

In the past, the tribal people[22] were considered primitive, ferocious and dangerous. They were regarded as brutal head hunters. The advent of Christianity brought about a great change in their attitude to life.

By listening to the preachings of the missionaries, the tribals developed a new outlook. For the tribal to become a Christian was to become a man more refined and polished.

The Christian missionaries and pastor teacher taught the tribal how to read and write. The missionaries taught them to abstain from intoxicating drink. Abstaining from drinking especially the traditional *Rice beer* led to decrease of social get together. This led to spending of more time with families.

---

* The Kuki included the Paite tribe till 1956-57.

They started wearing proper dress and understand the value of cleanliness. They realised the value of education. Personal cleanliness, cleanliness in the house and better ways of life were taught by the missionaries and the Evangelists which the tribals started to follow.

If a person fell sick the non-believer believed that he could be cured by sacrifices of birds and animals. This was a costly affair specially for poor persons. When they became Christians, all kinds of sacrifices became unnecessary.

The new faith brought peace among them. The tribals found in Chirstianity, a religion that met their needs which sets them free frm the bondage of evil spirits.

Their animistic religion did not teach them about sin and salvation. Inter-tribal war subsided. A new Christian society based on equality and mutual respect was established.[23]

Social Life

Christianity has a far[24] reaching influence on the *Paite* tribe. Community festivals and taboos and ceremonies were replaced by Christian festivals such as *New Year's Day, Good Friday, Easter Sunday* etc. There is total disappearence of animal sacrifices which used to consume a considerable portion of all the earnings of the tribal people.

Conversion from their animistic belief to Christianity, improved the Paite's social life. They discarded their old customs and practices. The dormitories or separate sleeping houses for boys and girls were no more necessary. According to their old custom, they used to bury the dead in front of the house. But now, every Christian village has a cemetry outside the village.

The practice of food offering to the spirits of their dead which they believed remained on full moon in the house was totally given up then since they believed that human spirits do not roam on the earth. Individuals and families helped each other in times of death and other misfortune in form of cash or deeds and also celebrate together during times of marriage, birth etc.[25]

Marriage and Family

With the preachings of the Christian missionaries the usual practice of marriage arrangements by parents is gradually disappearing and the girl's and boy's consent are becoming a deciding factor. The part played by *Zu* (rice beer) which is offered during marriage arrangements is replaced by tea.

Marriage is solemnised in the Church by a pastor with the bride wearing western style white gown, with the bridegroom having a bestman and the bride having a bridesmaid beside her have become a common practice.

Christianity taught the Paite tribe to realise that Christian life in one of its important facets means corporate living, corporate search, and corporate discovery. The age of marrying girl, was increased, before girls were married off as soon as they reach puberty.[26]

In the Paite family[27], of pre-Christian culture, the husband took pride in divorce and marring another. Such act was looked upon, by men as a sign of superiority and popularity in society.

Wives were considered the property of their husbands. Great changes were noticed in the family when Christ came into the hearts of a couple and brought the spirit of Christian marriage. They could both appreciate their respective positions in the light of the Gospel and as such, their new life became richer and meaningful in case of family and social life as well.

The relations between the parents and their children is that of love, respect and obedience.

The spirit of the Christian family and marriage is love which does not regard the wife consciously or unconsciously, as simply the one who looks his meal, washes his clothes, cleanses his house and brings up his children but is an equal partner in all the household matters.

The importance of education is realised by the parents, who send their children to schools and colleges. Daughters and sons are treated equally by parents. Women were gaining respect and many of them have started working or started their own business.[28]

Changes in Marriage is also Taking Place, Observations by Karin Kapadia among the Tamils

Among the Tamils[29], there is a preference for close-kin marriage, a woman retained comparatively high status and considerable degree of independence, because of the equal relationship that existed between her and the groom's family. The three categories of marriage was preferred earlier:

(i) Marriage with MB
(ii) Marriage with MBS
(iii) Marriage with FZS

It is the matrilateral kin who occupy centre-stage in non-Brahmin Tamil kinship because they are the most preferred affines, despite, the fact that inheritance is in male line. The unity of the kindred and the importance of *Sondom* (the 'Kindred', 'Kinship') and sees marriage as reproducing the existing links of *Sondom.*

But, now, things are changing. Marriage is now taking place in a 'Modern Manner', for money and status rather than 'For love of kin'. Marriage with rich VERRU (Strangers, Non-kin) is becoming prevalent.

The above changes are, because, with education and economic opportunities. The men and women are moving away from low-paid agricultural labour, and getting jobs which pay them more, like Clerk, Teachers, etc. They demand educated and well-to-do spouses.[30]

Worship and Literacy

The conversion[31] of tribal people to Christianity placed themselves in a peculiar predicament so far as their songs and dances were concerned. The traditional tribal songs and dances were replaced to some extent by Christian songs in western tune, though the tribal songs and dances are still very much prevalent and are performed during festive ceremonies.

The Paites have in the church, *Preacher* and other *missionaries,* who is supported by the church members offerings. The youth and the women members in the church visit the poor and needy, they go to console the members of the person's family who dies, visits the sick etc. The youth also organise meetings, get-together, learn songs and have prayer sessions together.

Every Sunday the members of the church come together to worship God and share with each other the teachings of the Bible. The Paite had no written literature in their language until the advent of Christianity in Manipur. The introduction of written language in the Roman script was great factor in the spread of the Paite literature.

Sanitation and Hygiene

The work of the missionaries in changing and developing the lifestyles of the Paites has been significant. Everyman preferred to sleep at his own house, before, they use to sleep in groups in dormitories where certain diseases can be easily communicated.

A mother's chewing of food for their babies and unhealthy mouth-to-mouth feeding was replaced by special cooking for babies and spoon-feeding. The practice of having toilet attached to the house and letting refuse freely for pigs, dogs and chicken in the open ground below was stopped.

A separate construction at some distance from the house with a simple septic tank is preferred now. Smoking has been inherent in the blood of the Paites from generation to generation, especially smoking from a smoking pipe made of bamboo or the practice of drinking *tui buk* (a drink made from bamboo leaves).[32]

But now, all these have been given up to a large extent. Washing their hands before eating, praying before each meal, eating in separate plates, brushing teeth, wearing clothes which are comfortable and mainly western in style like trousers, shirts, skirts, jeans etc. are common now.

Economic Life and Other Changes

Because of the work undertaken by the Christian missionaries in establishing schools, colleges and other learning institutions, the Paites have become mobile in the sphere of occupations. They are no longer just farmers, cultivators, but many of them have gone to become doctors, engineers, administrators etc.

Girls have gained respectability and they are no longer just housewives. They have taken up jobs like teaching, nursing etc. Consumption of material goods and purchasing power of the Paites have increased manifold.

Traditional goods have given way to western goods like chairs, tables, umbrellas, modern aluminium and other utensils. Changes have taken place in the matter of food habits as well, as many items of fresh vegetables, fruits, pulses and the like are added to the otherwise restricted diet of rice and meat.

New methods of farming and cultivation have emerged. Jhuming cultivation has given way to use of fertilisers, tractors, pesticides etc. to increase the yield of crops.

Foodgrains are no longer wasted in unnecessary feasts or celebration but properly stored. The hard earned money is not spent on lavish things but invested in banks for future purposes.

Houses are built with proper ventilation and drainage facilities. Household items like T.V., Fridge, Telephones are now a common feature among the Paites.

NOTES

1. Emile Durkheim, *Elementary Forms of Religious Life* (1912), p. 47.
2. In, Madan and Majumdar, *An Introduction to Social Anthropology,* (1986).
3. *Ibid.*
4. Marx's in M. Haralambos, *Sociology: Themes and Perspective,* (1980), p. 460.
5. Emile Durkheim, *Elementary Forms of Religious Life,* (1912).
6. *Ibid.*
7. A.R. Radcliffe Brown, *Structure and Function in Primitive Society,* (1952).
8. *Ibid.*

9. T.N. Madan in, *The Journal of Asian Studies,* Vol. 46, No. 4, November, 1987, p. 748.
10. Baveja, J.D., *The Land Where the Bamboo Flowers,* (1970).
11. *Ibid.*
12. *International Encyclopaedia of Social Sciences,* p. 365, Vol. XIV.
13. *Ibid.*
14. T.B. Botto More, *Sociology,* (1964).
15. *Ibid.*
16. S.C. Dube, *Modernization and Development, (The search for alternative paradigms),* 1983.
17. *Ibid.*
18. S.C. Sharma, *Insurgency or Ethnic Conflict,* (2000).
19. *Ibid.*
20. S.A. Ansari, *Manipur Tribal Demography and Socio-Economic Development,* (1991).
21. *Ibid.*
22. S.A. Ansari, *op. cit.* No. 20.
23. *Ibid.*
24. T. Liankhohau, *The Paite: A Transformed Community,* (1994).
25. *Ibid.*
26. T. Liankhohau, *op. cit.* No. 24, pp. 140, 147.
27. *Ibid.*, p. 140.
28. *Ibid.*
29. Contribution to Indian-Sociology 27, 1 (1993), pp. 26-51.
30. *Ibid.*
31. T. Liankhohau, *op. cit.* No. 24, pp. 141-143.
32. *Ibid.*

6

Conclusion

The *Paite* tribe has undergone what *Srinivas* had termed as *"Westernisation"* which is defined by him as change brought about in Indian society and culture as a result of *150 years* of *British rule.* The coming of the Britishers especially the *British missionaries* has contributed to the evolvement of what the *Paite* tribe are today.

The education that took place through the missionaries established schools and learning centres has helped the *Paites* to be polished in their ways of life. The *Paite* society is more or less an egalitarian society, there is no caste system existing. There is not much of a rich or poor divide.

The *Paite* tribe are today, recognised as one of the *Scheduled Tribe of India,* which came into enforcement in the year *1957*, when a list of tribes was drawn out by the *Government of India.*

This recognition entitles the *Paites* of protection and of their progress, development and well-being. The *Paites* have established themselves as a distinct tribe having their own customs and traditions. The other tribes of the hilly regions of *Manipur* have more or less similar way of life, the only difference at times will be each tribe's own distinct dialect.

It will be of interest to examine the relation of the hill tribe with the mainstream society, in this case it will be the *Meiteis* who mainly inhabit the plains of *Manipur* and are in majority. The *Meiteis*[1] are one of the few *Indo-Mongoloids* who possesses a written script and a fairly rich literature from an early age. They have a sense of history and most of their manuscripts are historical records.

The history of *Manipur* is undoubtedly the history of the *Meiteis*—the *ruling community of the State.* From the most credible tradition we find that the Meitei community was a product of an amalgamation of not less than *seven communities. Meiteis* means *Mei*—People, *Tei*—Different.

Nestled between *Assam* and *Burma* (now Myanmar), the little state of *Manipur* had been for centuries the home of several ethnic groups. From the time immemorial, the *central valley of Manipur,* had been the home of the *Meiteis* while the *surrounding hilly regions of the State are inhabited by the Nagas and other tribal communities.*

Hinduism came to *Manipur* in the *18th century,* which became the state religion. The tribal population in the hills remained more or less influenced by the process of mass *proselytism of the Meiteis into Vaishnavism.* About a century later, *i.e.* the latter part of the *19th century,* Christianity began to enter the tribal belt of Manipur and began its conversion of the tribal population into Christianity.

A major example of social distance prevalent since the *Pre-Hindu to Hindu* period was between the *Meiteis* of the valley and the tribals of the hills. Cultural and linguistic difference coupled with geographical isolation and political condition between the two groups were responsible for the distance which became wider than ever when the *Meiteis* became *Hindus.*

To the *Orthodox Meitei Hindus,* the tribals were considered like the untouchable Hindus and no tribal was allowed to enter the house of a Meitei Hindu. For that matter, the Manipuri Muslims too were treated in the same vein by the Orthodox Hindu Meitei.[2] Though, culturally different, the *Meiteis* and *Paites* have been living alongside with each other peacefully.

The *Meiteis* have accepted the *Paite* tribe and the other tribes as one of the significant communities in the society. The tribes are giving freedom to follow their rules, customs and traditions and their lands are not disturbed by the mainstream people (here, it is the Meiteis).

Marx, Weber, Mannheim and several other sociologists have been continuously preoccupied with the study of their own societies. It is evident however that a sociologist engaged in the study of his own society enjoys advantages as well as disadvantages and it is very important that disadvantages are minimised while the advantages are retained.

The sociologist[3] who is engaged in the study of his own society is likely to be influenced by his social position, not only in his observations but also in the problems he selects for study. In order to observe any society, the observer needs a measure of detachment from his own and for detachment to be effective, it must be as much a matter of the emotions as the intellect.

The seminars and lectures, the informal discussion outside the classroom and lecture hall and most of all the very processes of writing about his experience for an impersonal and professional audience, gradually produce for him a measure of distance from the field he left behind.

Physical distance from the field, as well as the necessity of describing and analyzing his experiences in terms that will be intelligible to his colleagues all over the world, forces him to emerge from his previous role of participant observer and become an impersonal catalyst.[4]

The *Paite* tribe displays instances of mechanical solidarity in the past as written about by *Durkheim.* What *Durkheim* meant by mechanical solidarity is that there is a strong system of social links between the individual and society based on obligation, custom and sentiment. Collective rules and social practices are predominantly religious in nature and a strong common conscience pervades all aspects of social life.

The kinship group is the dominant social institution. The population is homogenous small and isolated. There is a low degree of individual autonomy. There is a system of beliefs which is uniformly diffused throughout the society creating uniformity in attitudes and actions.[5]

Likewise, the *Paite* tribe during the past had to a large extent, these characteristics of mechanical solidarity. The village and its people were strongly together. The population was small and the economy was mainly for self-consumption. People helped each other in cultivating each others lands and punishments and sanctions were based on the tribal laws and customs or as decided by the village chief.

Now, at present, mechanical solidarity has given way to *Organic solidarity.* When *Durkheim* wrote about organic solidarity, he meant, occupational functions becoming more specialized. There is larger populations spread over broader geographic areas.

There is increased autonomy of the individual based on legal rights and political freedom system of social relations in which individuals are linked by contacts rather than sentiment or obligation.

Likewise, the *Paite* tribe in present time is showing some characteristics of organic solidarity. The individual no longer is very much bounded by village rules or religious beliefs. He has started setting up his own nuclear family, moving away from joints of family network. He starts gaining specialization in various fields.

Occupations like Doctor, Nursing, Engineering, Lecturership, Administrative Services etc. become popular, shifting away from the traditional occupation of cultivating and farming.

The *Paite* tribe also displays what *Radcliffe-Brown (in his edited book, African systems of Kinship and Marriage)* has written about the *three principles,* at least *one of the principles exist among the Paite, i.e.* unity of the sibling groups.

The principle of Unity of the *Sibling* group refers to the united nature, the solidarity and togetherness of the Sibling group as viewed by an outsider. Likewise, the *Paite* tribe also displays such principle as Sibling helped each other in times of need and other crisis.

In a classificatory system, a man uses the same term for his father and his father's brother because he thinks of them as relatives of one general kind. The principle on which the classificatory terminology is based may be called the principle of unity of the sibling group. This principle of unity of the Sibling group exists among the Paite tribe.

The other Two Principles which Radcliffe-Brown talked about are: *Generation Principle* and *Unity of the Lineage Group.*

Generation Principle

Between two proximate generations, this relation is normally one of essential in equality, authority and protective, respect and dependence. But between, the two generations of grandparents and grandchildren, the relation is a contrasting one of friendly familiarity and near equality.

Unity of Lineage Group

This feature exists among the *Shona* tribe, an *African* tribe, all the male members of the wife's lineage through all generations are called *'father-in-law'*. All the female members of a man's mother's lineage in her own and succeeding generations are his mothers.

The *Paites* are a tribe in transition. They embody both traditional as well as modern values. They retain traditional customs like appointing *Inndongta* (Household Council), for marriage, the boy's family go to the girl's family with a kettle of tea, feasting together at the time of marriage, naming of a child is by taking the last syllable of the person who is naming the child, all are still practiced by most of the *Paites,* at the same time, modern lifestyles have become quite prominent.

People are moving out of their villages and settling down in towns and cities. Nuclear familes are being set up, new occupations like engineering service, medicine, law, management, administrative service have been taken up, moving away from agricultural activities. The authority of the chief has declined, now democratically elected institutions has emerged.

But in recent times, the region of Manipur[6] has also witnessed inter-ethnic-conflicts often backed by their respective insurgency groups. The emergency of ethnic assertion and contest for social and political control pose serious political problems and thus needs serious consideration.

This assertation is based on "fears" of assimilation especially by the dominant groups and the need at the same time to protect their ethnic territoriality and natural resources, *i.e.* land.

Ethnic conflicts in Manipur has been intensed as witnessed between the Kukis and Nagas, Kukis and Paites, Meiteis and Pangas (Muslim).

Today, a very serious problem faced by Manipur could be observed by the following reasons—

(i) All social groups manifest distinctive social and cultural personality and linguistic entity.
(ii) There is no cultural and social organisation and interaction despite their physical affinity.
(iii) Religion factors, there are Christians, Hindus and Muslims.

Today, concepts of *"Us"* and *"Them", "Dominant"* and *"Minority"* and *Tribal and Non-Tribal* are very much prevalent.

Manipur has many ethnic groups who resemble each other physically have similar traditions and customs where the only dividing factor is that they speak different dialects.[7]

In Manipur, ethnicity has replaced the idea of community, ethnic groups having loyal affiliation only to their groups has emerged. The idea of ethnicity is exploited by politicians and others for personal gains rather than little to do with well-being of the community.

In recent times[8] the Paites are becoming more and more conscious of their ethnic identity. Let us define what ethnicity is all about. The members of an ethnic group are socially integrated and united by an emotional bond. An ethnic group refers to an endogamous formation. Which is well bounded by its members who identify themselves distinctly from others. It is a social category of people who have a common culture that sets them apart from others in a society.

Ethnicity is not an inborn quality but grows out of sentimental and emotional bonds and common cultural heritage of people. Gordon (1976: 30) refers to ethnicity to a group with a shared feeling of peoplehood.

In the modern complex society, ethnicity is not the product of isolation. The persistence of ethnic groups can be carried on by interaction with other groups of the society. "We" feeling of a group may be dormant until the groups comes in contact with another group.[9]

It is somewhat amazing that in spite of the force of politics motivated only by power, devoid of any serious moral concern, of global trade, and commerce making inroad into the North East, of the growth of and expansion of modern science and its incredible technological spin-off, some of the cultures, traditions of the Paite tribe have retained a genuine sense of autonomy and a feeling that deep down their original ways of life are still alive.

NOTES

1. K.B. Singh's, *Social Stratification and Mobility in India,* (1977).
2. *Ibid.*
3. M.N. Srinivas, *Social Change in Modern India,* (1972).
4. *Ibid.*
5. Emile Durkheim, *The Elementary Forms of Religious Life,* (1912).
6. *Economic and Political Weekly,* September 29, 2001, pp. 3674-3677.
7. *Ibid.*
8. *Eastern Anthropologist,* July–September, 1991, pp. 227-239.
9. *Ibid.*

Bibliography

Books

Allen, B.C.: *Naga Hills and Manipur: Socio-Economic History,* Delhi, Gian Publications, 1980, Reprint of 1905 (Edited).

Ahluwalia, B.K.: *Social Change in Manipur,* Delhi, Cultural Publications, 1984.

Ansari, S.A.: *Manipur Tribal Demography and Socio and Economic Development,* Delhi, Daya Publishing House, 1991.

Beteille, Andre: *Society and Politics in India,* Oxford University Press, 1991.

Bertram, Carey S. and Tuck, H.N.: *The Chin Hills,* Vol. I, Tribal Research, Aizawl, Mizoram, 1896.

Bose, N.K.: *Culture and Society in India,* Bombay Publishers, Asia, 1977.

Bottomore, T.B.: *Sociology,* London: George Allen and Unwin Ltd., 1964.

Brown, R.: *Statistical Account of Manipur,* Delhi, Sanslaram Prakashak, 1975, Reprint of 1874 (Edited).

Das, K.K.: *Manipur Tribal Scene,* New Delhi, Inter India Publication, 1985.

Dube, S.C.: *Modernization and Development: The Search for Alternative Paradigms,* New Delhi, Vistaar Publications, 1983.

Durkheim, E.: *The Elementary Forms of Religious Life,* New York, Collier Books, 1961.

Epstein, A.L.: *Craft of Social Anthropology,* London, Tavistock Publications, 1967.

Fox, Robin: *Kinship and Marriage—An Anthropologist Perspective,* Harmondsworth, Penguin, 1967.

Furer-Haimendorf, C.V.: *Gonds of Andhra Pradesh,* Delhi Vikas Publishing House Pvt. Ltd., 1979.

Gougin, T.: *History of Zomi,* Manipur, Zomi Press, 1984.

Grieson, G.A.: *Linguistic Survey of India,* Vol. III, Part-III, Calcutta: Office of the Superintendent, Government Printing, India, 1903.

Guite, Piangzathang: *The Story of Paite in Indo-Myanmar,* 1995. (Unpublished)

Haralombos, M. and Heald, R.M.: *Sociology: Themes and Perspective,* London, Oxford University Press, 1980.

Jamkhanthang, T. and Thangchinlian, G.: *Paite Customary Laws,* Churachandpur, 1986. (Unpublished)

Johnstone, James: *Manipur and the Naga Hills,* (Previously Published as: 'My Experiences in Manipur and Naga Hills'), 1876.

Kamkhenthang, H.: *The Paite: A Transborder Tribe of India and Burma,* Delhi, Mittal Publications, 1988.

Karotemprel and Deepali Danda: *The Tribes of North East India,* Calcutta, Firma K.L.M., 1984.

Lahir, K. Dilip and Dev, B.J.: *Manipur Culture and Politics,* Delhi, Mittal Publications, 1987.

Leach, Edmund: *Social Anthropology,* New York, Oxford University Press, 1992.

Liankhohau, T.: *Social, Cultural, Economic and Religious Life of a Transformed Community: A Study of Paite Tribe,* New Delhi, Gyan Publishing House, 1993.

Linton, Ralph: *The Study of Man,* Appleton-Century Crofts, 1964.

Mate, Lansung D.: *The Tribe of Manipur: A Case Study of Twisom Jang,* New Delhi, Omsons Publications, 2000.

Mitra and Gupta: *A Common Perspective for North East India,* Delhi, Daya Publishing House, 1991.

Murdock, G.P.: *Social Structure,* New York, MacMillan, 1949.

Oberoi, Patricia: *Family, Kinship and Marriage in India,* Delhi, Oxford University Press, 1993.

Panchani, C.S.: *Manipur: Religion, Culture and Society,* Delhi, Konark Publishers, 1987.

Radcliffe-Brown, A.R.: *Structure and Function in Primitive Society,* London, Routledge and Kegan Paul, 1952.

Radcliffe-Brown, A.R. and Forde, Daryll: *An African System of Kinship and Marriage,* London, Oxford University Press, 1967.

Ray, Ashok Kumar: *Authority and Legitimacy: A Study of Kukis in Manipur,* Delhi, Renaissance Publishing House, 1990.

Reid, A.S.: *Chin-Lushai Land, Gazetteer of Bengal,* Delhi, Gyan Publications, 1980.

Roy, J.: *The History of Manipur,* Calcutta, East Light Book House, 1973.

Sen, Shipra: *Tribes and Castes of Manipur—Description and Bibliography,* Delhi, Mittal Publications, 1992.

Shakespeare, J.: *Lushei-Kuki Clans,* Delhi, Cultural Publishing House, 1983.

Sharma, S.C.: *Insurgency or Ethnic Conflict (with Reference to Manipur),* New Delhi, Magnum Business Associates, 2000.

Singh, K.B.: 'Social Stratification and Mobility in India from North East India', *A Sociological Study* edited by S.N. Dubey, Delhi Concept Publishing Company, 1977.

Singh, Yogendra: *Modernization of Indian Tradition: A Systematic Study of Social Change,* New Delhi, Rawat Publications, 1996.

Tiwari, S.K.: *Antiquity of Indian Tribes,* New Delhi, Swarup and Sons, 1998.

Walker, Anthony R.: *The Toda of South India,* Delhi, Hindustan Publishing Corporation, 1986.

Zehol, Lucy: *Ethnicity in Manipur, Experience, Issues and Perspectives (Diagram),* New Delhi, Regency Publications, 1998.

Articles and Journals

Barua, Indira-Minati Chutia: 'The Chaudang of Assam: Ethnicity and Social System' in *Eastern Anthropologist,* Vol. 44, No. 3, July-September, 1991, pp. 227-239.

Chin (Tribes) in Encyclopaedia Britannica, Vol. V, p. 555.

Family in International Encyclopaedia of Social Sciences, Vols. V and VI, pp. 301-322.

Kapadia, Karin: 'Marrying Money: Changing Preferenc Practice in Tamil Marriage in Contributions to *Sociology* (N.S.) 24, 1, 1990.

Parkin, Robert: 'Terminology and Alliance in India: Tribal S and the North-South Problem' in Contributions to *Sociology* (N.S.) 24, 1, 1990, pp. 61-76.

Shimray, V.A.: 'Ethnicity and Socio-Political Assertati Manipur Experience', in *Economic and Political W* September 29, 2001, pp. 3674-3677.

Social Change, in *International Encyclopaedia,* Vol. XI 365-375.

Virginius Xaxa: 'Tribes as Indigeneous People of India' in *Ecc and Political Weekly,* December 18, 1999, pp. 3589-3

Index